Educational Communications and Technology: Issues and Innovations

Series Editors
Michael Spector
M.J. Bishop
Dirk Ifenthaler

More information about this series at http://www.springer.com/series/11824

Ian A. Lubin
Editor

ICT-Supported Innovations in Small Countries and Developing Regions

Perspectives and Recommendations for International Education

Springer

Editor
Ian A. Lubin
Independent Scholar and Researcher
Savannah, GA, USA

Educational Communications and Technology: Issues and Innovations
ISBN 978-3-319-67656-2 ISBN 978-3-319-67657-9 (eBook)
DOI 10.1007/978-3-319-67657-9

Library of Congress Control Number: 2017952700

Printed on acid-free paper

This Springer imprint is published by Springer Nature
The registered company is Springer International Publishing AG
The registered company address is: Gewerbestrasse 11, 6330 Cham, Switzerland

Foreword

The most exigent needs in education surround its quality; yet, our steadfast and urgent response is to provide more access. The frenetic pace of change today is staggering in its breadth and scope. Two overarching dimensions characterize this change. The first is technology. The tools with which we extend our reach have become so sophisticated that we can scarcely ever understand our relationships with them. Their ability to help us compress time and transcend space has led to the emergence of the second overarching dimension of our era: globalization. Globalization is a process through which various forces minimize social, cultural, and economic distances and unmoor local influences, creating translocating phenomena that dominate an increasing number of spheres in our lives.

Technology makes globalization possible. The technology of transcontinental navigation enabled conquest and colonization on a scale unimaginable before the sixteenth century. In just over three centuries, the technology of some 20,000 transatlantic voyages, motivated by profit and the urge for power, created an African diaspora. This institution of slavery reverberates today. The term "human trafficking" conjures up poignant images of global movement in a world of inequality—all made possible through the application of particular technologies for particular purposes.

Globalization of communications has seen ICT applied to the problem of teaching and learning. International educational development organizations have put ICT at the forefront of their initiatives. The richer countries of the world have declared their commitment to *education for all* and to shared global prosperity and peace. We hope that ICT can provide access to education and e-commerce in order to benefit the world's marginalized, most isolated, and poorest. Thus, the United Nations has put forward a global education agenda, complete with prescribed policies and priorities for the developing world to embrace and adopt.

Despite these agendas and efforts, the world's peoples are not equally at risk. Nation states are not equal in natural and human resources nor in strategic vulnerabilities. They are also not equal in their susceptibilities to environmental damage, often wrought by the larger and wealthier nations. In this regard, *small* developing societies must be viewed as facing even greater challenges in communications and technology than the rest of the developing world. An attractive “solution,” frequently proffered, has been to counter the apparent deficiencies by increasing access to technologies, especially in educational settings.

However, mounting evidence indicates that providing access to technology does not necessarily improve the quality of education. In fact, while ICT has the potential to support education quality and to promote equity and equality, the smallest developing nations are seldom equipped to make the best use of the advantages of these borrowed technologies. Without critical and sensible implementation, these imported innovations also have the potential to further aggravate existing inequalities in impoverished contexts. Just like the imperialism that preceded it, globalization gives but takes away as well. We may gain more expanded access to the global community but lose local languages and histories, with entire cultures becoming homogenized.

This volume seeks to explore this complex set of paradoxical processes and the inherent inequalities of globalization and international education development, while particularly focusing on ICTs for learning in small states and developing regions. The authors of this volume are not claiming to solve these problems; rather, they point to dilemmas and contradictions in education that have emerged for small societies in the globalized world. The contributors argue that the pursuit of better quality in education through ICT is both nobler and more adequate a goal than simply increasing educational provision.

This book will be of interest to educational technology researchers, policymakers, and practitioners who are involved or interested in advancing educational communications and technology in international contexts. This could include but is not limited to educational technologists in private and public settings, administrators, teachers and faculty, technology researchers, instructional designers, and technology evaluators.

Mmantsetsa Marope, Ph.D.
Director, International Bureau of Education
UNESCO
Geneva, Switzerland

Acknowledgments

Many kind and generous people were involved in this project. The author would like to thank the following individuals for their comments and other assistance regarding this volume including Luanne Serieux-Lubin, Alvinus Melius, and Michael Thomas. A special thank you is extended to all the contributors and reviewers who have made this volume possible. Heartfelt appreciation goes out to Robert Kozma (happily retired), who provided thoughtful advice during the development of this book.

Contents

The original version of this book was revised. An erratum to this book can be found at DOI 10.1007/978-3-319-67657-9_9

Contributors

Tian Belawati Universitas Terbuka, Banten, Indonesia

J. Enrique Hinostroza Instituto de Informática Educativa, Universidad de La Frontera, Temuco, Chile

Michael Fahy School of Education, University of Michigan, Ann Arbor, MI, USA

Laura Hosman Arizona State University, Tempe, AZ, USA

Jeremiah H. Kalir School of Education and Human Development, University of Colorado Denver, Denver, CO, USA

Jeff Kupperman University of Michigan-Flint, Flint, MI, USA

Nancy W.Y. Law Faculty of Education, University of Hong Kong, Hong Kong, China

Ian A. Lubin Independent Scholar and Researcher, Savannah, GA, USA

Dewi A. Padmo Universitas Terbuka, Banten, Indonesia

Francesc Pedró UNESCO, Paris, France

Sdenka Zobeida Salas-Pilco Faculty of Education, University of Hong Kong, Hong Kong, China

Farrah M. Schiff Interactive Communications & Simulations, School of Education, University of Michigan, Ann Arbor, MI, USA

Jeff Stanzler School of Education, University of Michigan, Ann Arbor, MI, USA

Daniel A. Wagner International Literacy Institute, Graduate School of Education, University of Pennsylvania, Philadelphia, PA, USA

About the Editor

Ian A. Lubin is a professional educator and research scientist specializing in learning and instructional psychology, educational technologies, and international education development. He has served as an instructional faculty and researcher at various national and international institutions. Dr. Lubin is the author of several journal articles, chapters, working papers, and other consultation publications on the design of optimal learning environments, cognition and motivation for learning, utilization of instructional technologies, teacher/faculty professional development, and education for development in international contexts. Dr. Lubin serves as a specialist advisor to international development organizations and agencies such as UNESCO and the Commonwealth of Learning. He holds undergraduate degrees in Psychology and Multimedia Instructional Design and earned his M.Ed. and Ph.D. degrees in Instructional Psychology and Technology.

Part I
Introduction. Theoretical and Methodological Insights

Chapter 1
Global Forces, Local Needs, and ICT-Supported Innovations in Small Countries and Developing Regions: Overcoming the Misalignment. An Editorial Introduction

Ian A. Lubin

Abstract Developing countries receive international assistance allowing information and communication technologies (ICT) to be increasingly accessible for education development purposes. Despite substantial global investments to export technology-supported educational innovations, evidence pointing to improvements in education quality is mixed and inconclusive. This raises questions on how best to achieve implementation and adoption of technology innovations outside of the contexts that inspired their creation. Are the features of ICT innately and universally transferrable, supporting mass diffusion, or should we be paying greater attention to local social conditions that foster acceptance and use of ICT in developing countries? The unique contextual features of small developing countries impede their ability to take full advantage of these borrowed ICT innovations, even with relevant adaptations. Beyond mere access afforded by ICT, domestic issues of quality, equality, and equity in education are of growing concern for small countries and developing regions.

Keywords ICT educational innovations • Small countries • International development • Educational technology research

---

The original version of this chapter was revised. An erratum to this chapter can be found at DOI 10.1007/978-3-319-67657-9_9

I.A. Lubin (✉)
Independent Scholar and Researcher,
1305 Barnard St. # 81, Savannah, GA 31401, USA
e-mail: ianlubin@gmail.com

I.A. Lubin (ed.), *ICT-Supported Innovations in Small Countries and Developing Regions*, Educational Communications and Technology: Issues and Innovations,
DOI 10.1007/978-3-319-67657-9_1

1 Introduction

Many technology-supported educational innovations taking root around the world are driven by international development policies. These policies are fundamental to the global education agenda and depend heavily on the use of information and communication technologies (ICT). It is believed that ICT have the potential to improve learning outcomes while also promoting greater social and political equality and stability in a given society (see Kozma, 2005; World Bank Group, 2012). ICT can thus assist in generating the human capital needed for domestic social and economic growth as well as global competitiveness (World Economic Forum, 2015). These potentials of ICT are in part responsible for the set of *technology in education* priorities that are passed on to poorer countries by the richer countries of the world. These exported priorities are presented as gifts, commonly referenced as international assistance, however domestic reforms are expected in return. For developing countries, implementing foreign-based ICT-supported policy reforms can be financially burdensome and can also present unintended sociocultural and political consequences, such as creating new economic and social disparities among citizens (Di Battista, Dutta, Geiger, & Lanvin, 2015).

In general, despite the many promises, it remains a matter of debate whether ICT innovations for international development are worth the substantial investment. For example, while nearly 75% of the World Bank's project portfolio contains ICT components, results are reported to be mixed, with less than 60% of projects reaching or likely to reach their objectives (World Bank Group, 2012). Additionally, the World Bank's Independent Evaluation Group has concluded that after more than USD 4 billion in investments in ICT, project outcomes have regularly been unsuccessful (Independent Evaluation Group, 2011).

A fair amount of education research and evaluation is focused on a separate question concerning whether ICT can actually improve learning outcomes and the quality of education. The results are mostly inconsistent and inconclusive (Clarke, Wylie, & Zomer, 2013; Lowther, Inan, Strahl, & Ross, 2008; Lowther, Ross, & Morrison, 2003; Tamim, Bernard, Borokhovski, Abrami, & Schmid, 2011), or they point to outright failures (Dodson, Sterling, & Bennett, 2012). For further discussions see Higgins, Xiao, and Katsipataki (2012), Kozma and Wagner (2005), and Winthrop and Smith (2012). In spite of the alarming rate of implementation and adoption failure and the inconclusive evidence of ICT's impact on education quality, the international donor community and local beneficiaries in the developing world appear to be engaged in ongoing policy endeavors geared toward the continued use of ICT for educational improvement and development. Developed countries see ICT as an attractive option for modernizing education in poorer countries. Developing countries view the inability to take advantage of the ICT revolution as being left behind in a globally competitive world.

The policies mentioned above reveal a broader thematic agenda for education that touts the utopian goals of globalization: poverty reduction, shared prosperity, open borders and markets, equality of opportunity, and a global ubiquitous culture.

Yet, with globalization, the questions must always be raised about what local knowledge, cultures, languages, identities, and ways of being in the world are being homogenized out of existence. Furthermore, we must question whether and how we, in the field of ICT for learning and development, are being the arbiters of the extermination.

While a comprehensive review of the impacts of globalization on education is beyond the scope of this chapter, there are a few points to consider here that are pertinent to a more balanced conversation about the implementation of ICT internationally. First, the transfer of international priorities (diffusion of knowledge) mentioned above is not unidirectional, that is it does not proceed solely from advanced countries into the developing world. Secondly, foreign-based policy reforms are regularly contested and even sometimes rejected in the developing world. Consequently, it is important not to rationalize developing countries as impotent, inert beneficiaries of exported policies. It is equally important not to misattribute the failures of ICT implementation as stemming strictly from the hegemony of globalization forces. In fact, a wide variety of contextual contingencies in developing countries seems to suggest that technology innovations should not be expected to work in the developing world as well as they do in their places of origin. Consider instead that international education policies and programs go through many transformations as they are repackaged to fit domestic level needs in their places of implementation. Also, this process of reshaping global priorities in local spaces, in turn, exerts influence on how policies are re-conceptualized and re-constituted in their places of origin.

Lacking in our research endeavors are more opportunities to assess the role of local contexts: the social conditions, governance, politics, histories, and ecologies that may influence the implementation and adoption of the global ICT in education agenda. Current research, for instance, does not examine the local development needs and ambitions of small countries and developing regions (SCDRs) to determine how, if at all, they may shape or reshape the thematic agenda for international education and development involving the use of ICT.

This volume seeks to illuminate processes, transactions, discourses, and learning experiences in small developing countries as they relate to technology-supported innovations in education. It examines how the implementation of educational technology internationally has created a particular attitude toward knowledge diffusion that is often uncritical and culturally misaligned. Notwithstanding, the volume highlights the ways in which the current ICT landscape is being mutually shaped by the interrelations among international donor and research communities and the relevant actors in local places of implementation. The project brings together technologists, researchers, scholars, academics, and industry professionals with a range of expertise in applying technology to educational needs in developing contexts. The chapters that follow reveal the diversity of the issues and also of the contexts. Small developing countries have common characteristics but are remarkably heterogeneous in terms of cultural heritage, languages, identities, and behaviors.

The volume, therefore, is about educational technology trends and innovations as they transpire in small countries and developing regions: though, also by necessity, it draws from the field of education development, with elements of international and comparative education. This interdisciplinary approach has implications for how the various methods within the volume are discussed and how the potentials and special qualities of the inquiries featured in the text are presented. This also influences how the literature related to ICT is traced through social science and educational technology research. This opening chapter serves to set the mood, to introduce some historical and contemporary issues that are key to subsequent discussions, and to pay a proper acknowledgment to the countries and regions that are the local settings of the book as well as highlight the variety of challenges for ICT in education in these contexts.

2 Small Countries and Developing Regions

Developing countries are generally characterized as having less advanced industrial bases and persistently smaller economies than developed countries. In addition, developing countries are considered to have populations with lower levels of education, life expectancy, and household income. A distinctive set of developing countries, however, has an additional feature that further compromises their chances for advancement – that is their *small* populations and land area (keeping in mind that some larger countries can have small populations).

Small countries have traditionally been overlooked in educational technology research mainly because of their smaller population sizes and low-income statuses and also because ICT use in these contexts has generally lagged behind the more developed world. Yet small countries represent a sizeable portion of the world's independent states and studying them offers a unique and important opportunity to gain a comparative perspective into educational technology policies and practices.

There exist no formal criteria for defining small countries or states. Classifications are varied and new conceptions continue to emerge. In terms of a working classification for small countries, the literature points to states with an upper population limit of five million (e.g. Bacchus, 2008). The Commonwealth and the World Bank both use 1.5 million as the population marker for small states (World Bank/Commonwealth, 2000). Small *island* states feature prominently based on these classifications as they make up nearly 30% of countries with populations of fewer than five million. Importantly, some islands combine to form multi-island nations (e.g., The Federated States of Micronesia) while some others are grouped together (e.g. the Caribbean) because they share linguistic, cultural, and socioeconomic traditions. This volume, therefore, makes references to small countries as well as developing regions.

Even if countries may not meet the population criterion (i.e., they may have relatively larger populations), they may still be considered small due to economic, sociocultural, and other development factors (Crowards, 2002). While the population and land area are common proxies for a country's 'size', a way of judging a country's level of development is examining the status of its economy. For example, the International Monetary Fund (IMF) has classified the world's countries into what it calls the advanced economies versus the emerging and developing economies (IMF, 2015). A small country typically is considered the latter, as measured by the gross domestic product (GDP), gross national product (GNP), and human development index (HDI) as well as the standard of living and the industrial base. However, it is worth noting that not all small countries are considered as poor or having low-income economies – further complicating any attempts at concrete classifications.

Granting that there is no explicit way to define *smallness*, the World Trade Organization (WTO) has followed conventions set up by other international organizations mentioned above. Nonetheless, a notable contribution by the WTO is in categorizing small states in terms of their trade flows, more specifically their share of global trade export (Jansen, 2004; Nottage, 2014; WTO, 2002). This trade flow conceptualization brings additional nuance into describing this diverse set of countries but does not solve the problem of finding a discrete definition of small countries.

A group of territories from around the world has been selected as the illustrative subject of this volume. While not implying any notions of homogeneity, inclusion in this set corresponds to having all or a combination of the following characteristics: political independence and sovereignty, a small population size of less than five million, and a small or emerging economy as defined by development organizations. Most of these countries also have relatively small land areas.

2.1 *The Big Deal with Being Small*

Problems experienced in small countries are not altogether different from those experienced in other developing countries. For example, developing countries that are not small may similarly experience poor economic outcomes (Easterly & Kraay, 2000). So why should special attention be paid to small countries? Small countries provide a sensitizing context for understanding the people, practices, and rationales surrounding international development. The *work* of transferring international development policies and practices accounts for the concentrations of foreign non-governmental organizations (NGOs), donor agencies, researchers, technologists, scholars, and other development professionals present in these local contexts. Given that development problems are thought to be acute in small countries, due to the interplay of some of their unique features, and since problems and solutions are often better illustrated under aggravated circumstances, small countries could serve as the workspace as well as the magnifying lens for conducting ICT research and evaluation. In this volume, the chapters that follow all deal with particular ways that the diffusion of ICT has transpired among various international and local actors. The stories from small states contain lessons about contextual complexities that have exacerbated the potentials for ICT failure. Looking exclusively at development markers like the GDP or GNP of these states could mask their true economic vulnerabilities that also come as a result of contextual factors such as small size, remoteness, insularity, and susceptibility to natural disasters (Briguglio, 1995). Some of these unique features of small states that have implications for ICT implementation and adoption are discussed below.

Geography, land area, and population size Most small countries are either single- or multi-island nations scattered throughout a wide expanse of ocean, which means that they are isolated and insulated (World Bank/Commonwealth, 2000). On one hand, this isolation often creates diversity in local knowledge, languages, and cultures. Still, geographic location can impede development capacity in small territories in many important ways (Everest-Phillips, 2014). An example would be the manner in which remoteness and isolation impose higher costs for doing business, especially with regard to trade and transport. Small states are geographically removed from major markets making international trade costs prohibitive. Additionally, inter-island transport essential for agriculture, fisheries, and tourism is often irregular and unreliable (UNCTAD, 2014). ICT can reduce the costs of both regional and international trade, yet ironically, it is in these small countries that ICT is most underutilized.

Small land area and population sizes are associated with higher costs per unit for both public administration and physical infrastructure (World Bank/Commonwealth, 2000). Small countries tend to have public sectors that are relatively large as compared to larger developing countries, which can equate to higher costs for providing public services per person. According to the OAS Executive Secretariat for Integral Development, this high cost is the result of resource constraints in both human and physical capital (OAS/SEDI, 1997). Human capital includes the knowledge, skills, and other intangible assets held by the people of a country (Becker, 2008). While there is always a need for improvement of human capital, this is particularly important in the case of small countries because they lose many of their brightest minds and most skilled individuals to migration. Everest-Phillips (2014) argues, "The percentage of college graduates who migrate is as high as 86% in Guyana and 83% in Jamaica" (p. 8). This leaves these countries with "a small manpower base from which to draw experienced and efficient administrators" (Briguglio, 1995, p. 1617). Physical capital includes commodities such as machinery, buildings, and computers. In many instances, these resources are outdated and function inefficiently, at best, and the cost of providing and maintaining these resources can be unsustainable for small countries.

The small size of domestic markets Small market size directly affects the cost of accessing public utility services (e.g., the Internet) and makes it difficult to regulate such services (Kirkman, 2002). Since small countries are unable to provide services on their own due to high costs, they must rely on foreign companies. However, using the size of the potential market as the foundation, foreign companies can argue that market shares are too small to allow for competition, thus bargaining for exclusive provider rights. Consequently, small market sizes increase the existence of monopolies, which tends to result in little or no competition (Briguglio, 1995). Therefore, costs remain high and quality of basic services less than adequate.

To compound the above issue, the small size of domestic markets limits the development of indigenous businesses that depend on high volumes of commercial activity to make successful starts, increase their scope, and become sustainable. Homegrown businesses are presumably more suited to provide for the local needs of their domestic markets. However, because of high failure rates of local businesses, goods and services must be sourced from external companies and countries. This requires small countries to source foreign currencies to pay for large import bills, which increases the national debt.

Undiversified economic bases and reliance on primary commodities In addition to the high import predicament, because of their size, small states have difficulties with export diversification. Reliance on primary commodities increases economic vulnerability as these commodities "have little or no value added to them" (UN-OHRLLS, 2016, para. 3). Furthermore, reliance on a narrow range of export products means that small economies remain vulnerable to market fluctuations – both in terms of price and demand. In order to be competitive, small countries usually require preferential agreements/treaties (OAS/SEDI, 1997) that can come with political pressures to secure their implementation (World Bank/Commonwealth, 2000).

Extreme vulnerability to external shocks External shocks are defined as "the unexpected and unpredictable events that affect a variable" (UN-OHRLLS, 2014, p. 12). Most small countries are unable to adequately deal with such external events. An example is the rapid inflation of prices of imported food products that can specifically pose problems for small countries. External shocks also wreak havoc on domestic sectors. As an illustration, many small states spend more than 30% of their foreign exchange annually on fossil fuel imports (UNCTAD, 2014). Reliance on fossil fuels for electricity means even small increases in international oil prices can severely cripple the local energy sector in a given small country. This would impact economic activity in both the public and private sectors (UN-OHRLLS, 2014).

Extreme vulnerability to natural disasters Many small countries are tiny islands or archipelagos and, as such, they are quite vulnerable to hurricanes, earthquake, tsunamis, etc. (Everest-Phillips, 2014). Hurricane Matthew of 2016 resulted in the death of nearly 900 people in Haiti and caused up to 90% destruction to some areas of the island (BBC News, 2016). More recently, hurricanes Irma and Maria of 2017 devastated several small countries in the Caribbean (The Guardian, 2017). Many islands have airports and cities in coastal areas that are critically affected by severe weather patterns. Other infrastructures such as telephone, road, and electricity networks are also compromised during natural disasters. This affects trade and transportation in general; it takes some time to repair and reconnect these networks after the disaster and with damaged infrastructure comes the cessation of economic activity, which imposes additional losses. Natural disasters can have devastating effects on undiversified economies, decimating their agricultural and tourism bases. This level of destruction of public infrastructure inevitably requires governments to seek aid and loans from international lenders. The inexorable truth is that natural disasters are a major factor in increasing the burden of debt on small countries.

Historical and sociocultural considerations The vast majority of small developing countries share a colonial history. The impact of colonization on small countries is far-reaching. Most colonization was agriculture-based, which means these territories served, for the most part, as remote plantations. This scenario continued to play out long after independence, with newly sovereign states continuing to trade with their former colonizers (Umemura, 2016) often with preferential treatment agreements. These trade relationships would eventually become either unprofitable for the colonizers or subject to free-trade laws. Many small countries were decolonized in the 1970s and some even more recently (Everest-Phillips, 2014). This implies that they are very young nations, still struggling to develop institutional capacity and continuing to build their private sectors. Political independence generally meant the termination of support and the withdrawal of key resources formerly provided by the colonizer. Ironically, in many cases, this political independence only reinforced the economic dependence on colonizers and on countries with more advanced economies.

Beyond economics, colonization also carried ethno-social implications for small states. In every case, colonization and discrimination were symbiotic processes. The colonized were always persons of a different race, religion, and/or culture from the colonizer. In the case of small islands both dependence and discrimination have continued and in many instances widespread inequalities have perpetuated ethnic rivalries, affecting the political and socio-economic development of these nations (Augustin, 2007).

Regional Integration Efforts As small countries develop and form socioeconomic linkages the need for geopolitical integration has become imperative. The concept of 'strength in numbers' suggests that acting as an individual state on certain issues is socially and economically detrimental. This causes some small states to attempt to strengthen their impact by seeking to interact with the international community as a regional body. Inclusive in this type of integration arrangement are regional policies that govern local policies and international interactions, thus reducing the autonomy of individual countries since decisions must benefit the entire regional body.

3 Classifying Small Developing Countries and Regions

The countries in this book are important because of their small size and their economic status as developing. However, it is easy to misrecognize these countries as being more homogenous than they are in reality. Doing this can have consequences for ICT in education endeavors. For instance, a landlocked territory in South America is different from a tiny island in the pacific, yet they are both considered developing. Failure to differentiate between these contexts more astutely means that a number of different social, economic, and ecological conditions may be conflated. Without more targeted classifications it is not likely that nuanced understandings of the role of technology can be properly cultivated. These distinctions are important in discovering why context matters in education, in general (Crossley, 2010), and especially as it relates to ICT.

Within the development literature there already exist classifications for developing countries. Below is some tabulation to bring into sharper focus the comparative significance of the smallness of countries appearing in this book. These comparisons are based on extant data and are derived by integrating multiple development indices from various development institutions. The classifications presented are not discrete or definite but they are intended to provide organizational guidance and help generate a better appreciation of the commonalities as well as differences among these countries.

3.1 Political Independence and Sovereignty

The first common characteristic among small, developing countries referenced in this book is their political independence and sovereignty. Table 1.1 includes sovereign countries with a population of fewer than 5M, as outlined in World Population Prospects (UN ESA, 2015). Many non-sovereign nations (e.g. Martinique, USVI, etc.) fit the population criterion but are excluded as these countries may have circumstances or realities that differ from fully independent countries. For example, Martinique is influenced by and benefits from being a present-day Department of France, in ways that Saint Lucia does not as a former British and French colony.

3.2 Comparing Income Categorization

Table 1.1 also illustrates other ways small developing countries are categorized, namely geographic region, population, and gross national income (GNI). Country populations are listed in millions. GNI range is listed in USD and is calculated using the World Bank Atlas Method (World Bank, 2017a). The figures listed for GNI are for 2015 and the groupings are as follows: low = under 1025; middle = 1026–12,475; and high = above 12,476.

3.3 Comparing Economic Base and Growth Potential

Yet another way to differentiate among small developing countries and regions is by grouping them based on the type of economy and potential for growth. In this regard, the physical, geographic, institutional, and social characteristics that determine their economic vulnerability and their relative position in the international community are examined. These characteristics influence the economy base and the

Table 1.1 Small states parameters: Region/Sub region, Population, Income Categorization (GNI range).

	Country Population in millions GNI: Low-income (L); Middle-income (M); High-income (H)		
Region	Less than 0.5 M	0.5–1.5 M	1.5 – 5 M
Africa			
	Seychelles (H); São Tome & Principe (M)	Comoros (L); Djibouti (M); Mauritius (M); Equatorial Guinea (M); Swaziland (M); Cabo Verde (M)	Central African Republic (L); Botswana (M); Lesotho (M); Namibia (M); The Gambia (L); Guinea-Bissau (L): Liberia (L); Mauritania (M)
Asia			
	Maldives (M): Brunei Darussalam (H)	Bhutan (M); Timor-Leste (M); Bahrain (H)	Mongolia (M); Armenia (M): Georgia (M); Kuwait (H); Oman (H); Qatar (H): State of Palestine (M)
Europe			
	–	Montenegro (M)	Republic of Moldova (M); Albania (M); Bosnia & Herzegovina (M); Croatia (H); TFYR Macedonia (M)
LAC			
Caribbean	Antigua & Barbuda (H); Bahamas (H); Barbados (H); Dominica (M); Grenada (M); Saint Kitts & Nevis (H); St. Lucia (M); St. Vincent & the Grenadines (M)	Trinidad &Tobago (H)	Jamaica (M)
Latin America	Belize (M)	Guyana (M); Suriname (M)	Costa Rica (M); Panama (M); Uruguay (H)
Oceania/Pacific			
Melanesia	Vanuatu (M)	Fiji (M); Solomon Islands (M)	
Micronesia	Kiribati (M); Marshall Islands (M); Micronesia (M); Nauru (H); Palau (M)	–	–
Polynesia	Samoa (M); Tonga (M); Tuvalu (M)	–	–

Sources: geographic regions (CIA, 2017), populations (CIA, 2017; UN ESA, 2015), GNI (World Bank, 2017a).

GDP composition by sector (i.e. the percent contribution of agriculture, industry, and services to total GDP). Some countries have agriculture-based economies, which mostly include farming, fishing, and forestry. Others have industry economies focusing on construction and manufacturing, mining, and energy production. Some have service economies involving activities that are not designed to yield production of a material good, such as finance, telecommunications, transportation, hospitality, and tourism (CIA, 2017). Table 1.2 presents these vulnerability classifications as conceptualized by the United Nations (UN-OHRLLS, 2017a, 2017b).

Small Island Developing States: SIDS Small Island Developing States (SIDS) are a group of developing countries from three geographic regions, namely the Caribbean; the Pacific; and the Atlantic, Indian Ocean, Mediterranean, and South China Sea (AIMS). These small islands face unique social, economic, and environmental vulnerabilities that negatively impact their sustainable development. These vulnerabilities include a narrow resource base and an inability to achieve economies of scale; small domestic markets; high costs for transportation, energy, infrastructure, communication, and services; high import bills; vulnerability to natural disasters; fragile natural environments; volatile economic growth; low and irregular volume of international travel; limited private sector growth; and high economic reliance on the public sector (UN-OHRLLS, 2011, 2017b).

Landlocked Developing Countries: LLDCs While some countries are susceptible due to their territorial openness, others are vulnerable due to their lack of access to the sea – having no coastlines or seaports. This is a different kind of isolation as they have no direct seaborne access to world markets and must depend on other bordering countries for trade routes. This invariably increases transit costs and the total cost of doing business. Being landlocked is one of the most difficult challenges to overcome and for that reason more than half of landlocked countries are also on the list of least developed states. The dependence on other countries for access to world markets makes LLDCs among the world's poorest (UN-OHRLLS, 2017a, 2017b).

Least developed Countries: LDCs The title of world's poorest belongs to the category of states called the least developed countries. As the title suggests, the extremely poor level of development puts LDCs at the bottom of the poverty scale with their contribution to global GDP at an astonishingly low 2%. This is the case even as LDCs have a combined population of nearly 900 million. In proper perspective, 12% of the world's population trades only 1% of the world's goods (UN-OHRLLS, 2011, 2017b). A cycle of poverty is the main constraint on economic growth and development. First, LDCs have agrarian economies and produce a small number of primary goods for export. This diminishes their abilities to attract or access foreign earnings. Then, the nature of the systems of production usually favors a small number of elite citizens. This is accompanied by the unequal distribution of income and political power across the populations. The human resource base is not well trained and so there is a lack of ability to access whatever scarce domestic resources are available. With little foreign earnings and attenuated domestic capac-

Table 1.2 Vulnerability classification: Small island developing states (SIDS), Least developed countries (LDCs), Landlocked developing countries (LLDCs). Economy base (GDP Composition): agriculture[1], industry[2], service[3]

Region	Vulnerability		
	Small Island Developing States (SIDS)	Least Developed Countries (LDCs)	Landlocked Developing Countries (LLDCs)
Africa	Cabo Verde[3], Comoros[1], Guinea-Bissau[1], Mauritius[3], São Tomé & Príncipe[3], Seychelles[3]	Central African Republic[1], Comoros[1], Djibouti[3], Equatorial Guinea[2], The Gambia[3], Guinea-Bissau[1], Lesotho[3], Liberia[3], Mauritania[3], São Tomé & Príncipe[3]	Botswana[3], Central African Republic[1], Lesotho[3], Swaziland[2]
Asia	Bahrain[3], Maldives[3], Timor-Leste[2]	Bhutan[3], Maldives[3], Timor-Leste[2]	Armenia[3], Bhutan[3], Mongolia[3]
Europe	–	–	Republic of Moldova[3], TFYR Macedonia[3]
Latin America & the Caribbean	Antigua & Barbuda[3], Bahamas[3], Barbados[3], Dominica[3], Grenada[3], Guyana[3], Jamaica[3], Saint Kitts & Nevis[3], Saint Lucia[3], Saint Vincent & the Grenadines[3], Suriname[2], Trinidad & Tobago[3]	–	–
Oceania/Pacific	Fiji[3], Kiribati[3], Marshall Islands[3], Micronesia[3], Nauru[3], Palau[3], Samoa[3], Solomon Islands[1], Tonga[3], Tuvalu[3], Vanuatu[3]	Kiribati[3], Samoa[3], Solomon Islands[1], Tuvalu[3], Vanuatu[3]	–

Source: Classifications (UN-OHRLLS, 2017a, 2017b).

ity, these countries seek out foreign financial assistance resulting in high debt that they lack the capacity to repay (UN-OHRLLS, 2011, 2017b).

Importantly, some small countries do not fit appropriately into the classifications listed in Table 1.2 and are excluded from the table including Albania[3], Belize[3], Bosnia & Herzegovina[3], Brunei Darussalam[2], Costa Rica[3], Croatia[3], Georgia[3], Kuwait[2], Montenegro[3]; Namibia[3]; Oman[2], Panama[3], Qatar[2], State of Palestine[1], and Uruguay[3].

3.4 Comparing Infrastructure and Capital

A further way of distinguishing among small developing countries is to look at the existing infrastructure per population. Based on the previous discussion on the value of both physical and human capital, Table 1.3 presents data relating to selected sources of capital among small developing countries. The table shows the actual population and the median age for each country (CIA, 2017; UN ESA, 2015); the percentage of the population that has access to electricity as per 2012 (World Bank, 2017b); the number of Internet users per 100 people as per 2014; and the mobile cellular subscriptions per 100 people (World Bank, 2017b). Finally, the table presents literacy rates in these countries, specifically for the percentage of the population that is age 15 and over who can read and write (CIA, 2017; UN UIS, 2017).

4 Aligning Local Needs with Global Priorities in ICT for Education

The push to implement educational technologies internationally corresponds with growing optimism about the use of ICT in the developing world to spur social and economic growth and to provide support to individuals and countries to get out of poverty (World Bank Group, 2012; Organisation for Economic Co-operation and Development (OECD), 2015). Even outside of formal education, it is hoped that computer literacy and Internet access coupled with the availability of computing devices, such as mobile phones and tablets, will impact the poorest of the world by expanding social interaction, knowledge sharing, and e-commerce (GSMA, 2008, 2013; West & Chew, 2014). Therefore, considerable investments have been made toward technology-supported educational reforms despite little evidence of the success of ICT in improving education.

There is also little evidence to suggest that ICT will be rejected or even demoted as a solution for improving economic conditions in poorer countries. Implementation decisions are frequently politically driven and technology-related rationales almost always allow for expediency. Connected to this anticipated reliance on ICT, this volume offers fresh opportunities to revisit the longstanding debates about how to achieve implementation and adoption of educational technology innovations in places other than where they were conceived and created. By concentrating on small developing countries, the volume seeks to understand what underlies the ongoing rationales for two main approaches: one that trusts in the innate transferability of technological affordances to any context, given appropriate adaptations; and another that sees the viability of technological innovations as resulting explicitly from the social conditions within beneficiary contexts. Understanding the implications for both approaches could contribute to refining the current knowledge base and to refocusing international efforts toward more successful results for ICT in education initiatives.

Table 1.3 Selected sources of physical and human capital in small countries

Country & region	Population	Median Pop. Age	Electricity penetration	Internet penetration	Cellular penetration	Literacy
Africa						
Botswana	2,262,000	24	53	19	167	89
Cabo Verde	521,000	25	71	40	122	88
Central African Republic, The	4,900,000	20	11	4	25	37
Comoros	788,000	20	69	7	51	78
Djibouti	888,000	24	53	11	32	N/A
Equatorial Guinea	845,000	21	66	19	66	95
Gambia	1,991,000	17	35	16	120	56
Guinea-Bissau	1,844,000	19	61	3	63	60
Lesotho	2,135,000	21	21	11	85	79
Liberia	4,503,000	19	9	5	73	48
Mauritania	4,068,000	20	100	11	94	52
Mauritius	1,273,000	35	100	41	132	91
Namibia	2,459,000	21	47	15	114	82
Sao Tome & Principe	190,000	19	61	24	65	74
Seychelles	96,000	33	100	54	162	92
Swaziland	1,287,000	21	42	27	72	88
Asia						
Armenia	3,018,000	35	100	46	116	100
Bahrain	1,377,000	30	98	91	173	96
Bhutan	775,000	27	76	34	82	65
Brunei Darussalam	423,000	31	76	69	110	96
Georgia	4,000,000	38	100	49	125	100
Kuwait	3,892,000	31	98	79	218	96
Maldives	364,000	26	100	49	189	99
Oman	4,491,000	29	98	70	158	91
Qatar	2,235,000	31	98	92	146	97
State of Palestine, The	4,668,000	19	98	57	78	99
Timor-Leste	1,185,000	19	42	1	119	68
Europe						
Albania	2,897,000	34	100	60	105	98
Bosnia & Herzegovina	3,810,000	42	100	61	91	99
Croatia	4,240,000	43	100	69	104	99
Montenegro	626,000	38	100	61	163	99

(continued)

Table 1.3 (continued)

Country & region	Population	Median Pop. Age	Electricity penetration	Internet penetration	Cellular penetration	Literacy
Republic of Moldova	4,069,000	36	100	47	108	99
TFYR Macedonia	2,078,000	38	100	68	106	98
Latin America / Caribbean						
Antigua & Barbuda	92,000	31	91	64	132	99
Bahamas	388,000	32	100	77	82	96
Barbados	284,000	39	91	77	129	100
Belize	359,000	24	100	39	51	83
Costa Rica	4,808,000	31	100	49	144	98
Dominica	73,000	33	93	63	127	94
Grenada	107,000	27	91	37	110	96
Guyana	767,000	25	80	37	71	89
Jamaica	2,793,000	29	93	41	107	89
Panama	3,929,000	29	91	45	158	95
St. Kitts & Nevis	56,000	34	91	65	119	98
St. Lucia	185,000	31	91	51	103	90
St. Vincent & the Grenadines	109,000	30	76	57	105	96
Suriname	543,000	29	100	40	171	96
Trinidad & Tobago	1,360,000	34	100	65	147	99
Uruguay	3,432,000	35	100	62	161	99
Oceania						
Fiji	892,000	28	59	42	99	94
Kiribati	112,000	22	59	12	17	N/A
Marshall Islands	53,000	23	59	17	29	99
Micronesia	104,000	22	59	30	30	89
Nauru	10,000	26	N/A	54	68	97
Palau	21,000	32	59	27	91	100
Samoa	193,000	21	100	21	56	99
Solomon Islands	584,000	20	23	9	66	84
Tonga	106,000	21	96	20	64	99
Tuvalu	10,000	25	45	43	38	N/A
Vanuatu	265,000	22	27	19	60	85

Sources: CIA Factbook (2017), UN ESA (2015), UN UIS (2017), World Bank (2017b)
Note: *N/A* not available

4.1 *Mass Diffusion of ICT for Education: International Transfer of Technological Innovation*

High levels of poverty and inequality in small (and big) developing regions have created a market for educational provision. In addition, the notion of *education for all* is acknowledged as a human right (Shileds, 2013). The World Conferences on Education for All, held in Jomtien in 1990 and Dakar in 2000, outlined the ambitions of richer countries to expand enrollment and participation in learning at all levels in poorer countries (UNESCO, 1990, 2000). Education strategies and objectives outlined in the UN Millennium Development Goals (MDGs) and subsequently the Sustainable Development Goals (SDGs) for post-2015 serve as further examples of continued efforts by educational stakeholders to advance a development agenda for education. Increasing numbers of educational products and services are being created to meet the demands of these mass markets in poorer countries.

The type of educational expansion mentioned above requires mass diffusion and adoption. In terms of ICT-supported innovations, mass diffusion is based on the reasoning that generic products created elsewhere are adaptable and transferable, in their original form, in recipient countries. It means the one-size-fits-all or the best practice approach[1] becomes inevitable. This has undoubtedly been the case with ICT innovations exported to small developing countries, and some important lessons are applicable. The first lesson is that mere access to technology in education does not guarantee the quality of that education (Kozma & Wagner, 2005). Several of the authors in this volume address this concern. The second lesson is that simply providing technology does not mean people want it or will use it (Rangaswamy & Cutrell, 2012). Yet, the donor community is reputed for imposing aid conditionality in exchange for economic support. For local governments and education stakeholders, the choice involves accepting and using technologies and organizational frameworks that are often not well-suited to their context or risk losing much needed financial and technical assistance. The third lesson involves the sponsored research and evaluation landscape that has now become favorably disposed toward quantitative-leaning methodologies e.g., impact evaluations of performance outcomes. This situation is partly the result of the donor community coming under increasing pressure to demonstrate efficiency to justify its substantial investments worldwide. These three lessons reveal that the challenges for mass diffusion of ICT innovations in education are as much related to implementation as they are to evaluation.

The early chapters of the book help to navigate the perplexing theoretical and research landscape of ICT for education development. In terms of the 'success' of technologies in improving learning, the authors stress the importance of identifying

[1] The best practice approach assumes that the desired result will be reliably achieved based on experience and research. However, much of the research and experience is from the developed countries that may have little in common with the developing countries to whom the *best practices* are exported. As such, best practices in one context may not be transferrable to another.

precisely what are the claims and what are the outcomes in developing countries that are being realized. In chapter two, Francesc Pedró examines the insufficiencies in the existing knowledge base on technology use in education. The author laments the state of affairs in which access to technology is prioritized over genuinely meaningful adoption. Pedró raises issues concerning the role of research in understanding how exactly technology is related to educational performance. Instead of relying on mass implementation, the author proposes a research agenda for the future that focuses on better monitoring and assessing of solutions that actually work, to better align them with the needs and realities of teachers, and to increase their scale accordingly. This will require modifications to current research strategies and even the nature of questions that are asked. Pedró believes the field must move beyond the question of whether or not to use technology and must now focus on which solutions have shown success and can feasibly be incorporated into sound teaching practices. This means that an evidence-driven knowledge base is essential in addressing the educational effects of technology and how it can be used.

In Chap. 3, Dan Wagner introduces an ICT for education (ICT4E) design framework and makes the case for the need for better design solutions that are measurable, sustainable, scalable, and culturally appropriate to support the Sustainable Developing Goals (SDGs). Wagner acknowledges that due to a great amount of publicity and buzz, ICT will continue to be promoted as a solution for improving the quality of education and development, on the whole. However, he asserts that claims of widespread success from ICT are premature and, in some cases, overstated. Wagner argues that we must be focused on what are the appropriate indicators for measuring and communicating successful learning outcomes. He believes ICT may hold legitimate value in improving the quality of education and life of persons in developing countries, as ICT have the reach and they offer tools that can bring about innovative solutions. However, in order to be meaningful, the purpose of ICT interventions must be well-defined if they are to contribute to what is known and how policy around educational technology is planned. Wagner cautions that sound policy planning and design solutions are important since ICT have the paradoxical potential to be both a vehicle to transform people's lives and an agent to grow the disparities between the rich and poor in developing countries.

4.2 Making Educational Provision a Matter of Quality, Equality, and Equitability

While it is important to provide access to technology for education, it is observed that uncritical provision can create even greater economic and social disparities (Di Battista et al., 2015; IEG, 2011). A more appropriate goal is to ensure that the distribution of technology-based education is equitable and promotes equality among members of society. Thus, education must be made uniformly available to all, including underrepresented individuals who are outside of mainstream educational attainment due to critical impediments - but this is certainly not an easy task. These impediments include diminished socioeconomic conditions and geographic isolation and/or displacement, all of which are commonplace in small developing countries.

An obstacle small developing countries face is an unfortunate internal perception problem. It is sometimes erroneously believed that international-grade education cannot be developed locally and must, therefore, be imported. Advanced economies reinforce this perception by suggesting that there are prescribed international quality standards in education that all countries must strive to meet to be globally competitive. The problem is how should citizens be trained to be highly skilled and educated in order to expand domestic human capital and, subsequently, how should these educated persons be retained against the dynamics of migration and brain drain. The solution involves creating policies on how to provide and market high-quality locally-developed education systems. For small countries, this is the consistent challenge of how to embrace the international and yet protect local interests. ICT are seen as indispensable to that effort as they allow individuals to stay home while having access to foreign and domestic education markets. Simultaneously, ICT also allow those who are geographically isolated to access education and be part of the economy.

A related and familiar policy issue involves the extent to which a government should be involved in providing education. The availability of and access to ICT-delivered education has generated a free market for education attainment and has forced new conceptions of students, not just as learners, but also as consumers and creators of education, with diverse options to gaining credentials. Education policies must now tackle the lingering issue of whether education should be a private commodity guided by market forces and driven by learner demands, or whether it should be a public good with obligations to be vested with domestic governments. In either case, questions of equality and equitability must be brought into sharper focus. Several chapters deal specifically with the development and implementation of ICT policies that guide participation, quality, and equity in education in small states and developing regions.

In Chap. 4, Sdenka Zobeida Salas-Pilco and Nancy Law artfully report on ICT curriculum planning and development in 16 small countries in the Latin America and Caribbean region. The authors discuss the above-mentioned issues from the perspective of curriculum planning for strategic development. While they recognize the need for all countries to respond to the way technology is changing education, the scholars suggest that curriculum planning for strategic development is more difficult for small, culturally diverse, and economically less developed countries. Salas-Pilco and Law determine these countries are not homogeneous and there is even tremendous cultural, linguistic, and ethno-social diversity within them. It is the very diversity of small states that poses additional burdens on already scarce resources needed to develop ICT-related curricula. The authors report specifically on three case studies of ICT curriculum planning and developing: Costa Rica, Suriname, and the Caribbean Universities Project for Integrated Distance Education (CUPIDE). They present lessons learned that could assist all who may be involved in developing and implementing ICT-supported curricula in small developing states.

In Chap. 5, J. Enrique Hinostroza takes a historical approach in looking at the evolution of ICT-in-education policies of developing countries over several decades. Hinostroza sifts through the complex policy landscape and reveals a new contemporary scenario where ICT are now ubiquitous in the developing world, to the extent that there is widespread use outside of school. Unfortunately, Hinostroza argues, this rapid availability of ICT is matched by increasing inequalities in educational provision, and is compounded by teachers' and students' lack of skills necessary to effectively use ICT. This is resulting in unsupervised, improper use of technology, outside of school settings, that has the potential to reinforce existing disparities. Hinostroza diligently outlines his case that widespread availability of ICT, without the proper professional supports and incentives for teachers, without due consideration to the ways students learn with technology outside of schools, and without strategies to marry students' learning with teaching abilities, can actually be detrimental to developing countries. The author makes special references to the country of Chile.

In Chap. 6, Dewi Padmo and Tian Belawati focus on the issue of sustainability of ICT-supported innovation policies. Their case study reports on the Universitas Terbuka, an open university in Indonesia, a multi-state nation comprising over 17,000 islands. The authors give an illuminating account of the history, development, and implementation of a government policy to increase equitability of access to education for citizens scattered throughout Indonesia, including those isolated in the poorest islands. The provision of ICT-supported open and distance education through Universitas Terbuka has transformed the educational landscape in the country and region. However, concerned about quality, equitability, and especially sustainability of open learning via ICT innovations, the institution's administrators have developed and implemented university-wide policies that govern the use of ICT. These policies undergird successful strategies for addressing local circumstances in regards to ICT for education.

4.3 International Partnerships Supporting Local Curricula Needs and Interests

Many small developing countries and regions recognize the need to adjust international goals and targets so that they are meaningful in their particular contexts. For instance, Caribbean country representatives openly express that "in order for the post-2015 development agenda to be relevant and avoid major programmatic and monitoring challenges, there must be a core set of universally agreed goals and targets with which the countries can identify at the regional and national levels with country and region specific indicators" (United Nations Economic Commission for Latin America and the Caribbean (ECLAC), 2015, p. 4). Countries sought "to take ownership of the SDGs…by establishing regional priorities for the Caribbean among the SDGs while ensuring that the goals are tailored to specific country development strategies" (p. 1). This was an effort to ensure that national and regional priorities were not sidelined in favor of multilateral and international goals and agreements with which they do not always align. Caribbean countries suggested that SDGs should also be transformed into "national level plans" that engage in a "participatory bottom-up community-driven development" (p. 6).

The desires expressed above by Caribbean countries signify what is now commonly referred to as contextualizing global policies to fit local needs. Many small developing countries have already created education policies that reflect their independence and sovereignty, histories, and cultural and national identities. This includes greater emphasis on the role of local knowledge and local curricula. Successful technology implementations under these circumstances require multi-stakeholder partnerships built on dialogue, cooperation, mutual participation, and recognition (Lubin, 2016). The final chapters of the book are examples of multi-year partnerships that provide more perspective on the challenges and rewards of culturally sensitive implementations in small developing countries and regions.

In Chap. 7, Jeremiah Kalir, Michael Fahy, Jeff Kupperman, Farah Schiff, and Jeff Stanzler examine the value of critical play in education, especially the notions of playful design and enactment of games for learning. The chapter describes the design and implementation of an ICT-supported, game-based learning innovation in the Sultanate of Oman, made possible through an international partnership. The authors argue that while attempting to provide access to ICT for improving learning in developing countries, we must be equally committed to also providing support for the types of innovative pedagogies and social practices that potentially could dismantle existing inequalities. This sociotechnical position is important given the tendencies for international development efforts to emphasize the provision of tools over associated practices. Further, they argue, the sociotechnical practices associated with play could be a viable approach to education development work in small countries in order to provide the type of twenty-first century learning that is needed in these contexts. The authors raise concerns about the proliferation of game-based learning as a form of ICT for development. As such, they advocate for an intentionally balanced approach that couples technologies with appropriate social and pedagogical practices.

In Chap. 8, Laura Hosman chronicles her development as a practitioner-scholar in the area of technology for schools in developing countries. Through a multi-sited autoethnography, Hosman narrates the stories of her experiences working in small countries, such as Haiti and the Federated States of Micronesia, which are among the world's poorest and have an extremely limited infrastructure to support ICT-enabled education. The journey begins with the author's decision to realign her professional focus from 'just' researching about ICT to becoming actively involved in implementing educational technology in remote rural schools in the developing world. Hosman describes how her teaching, research, scholarship, and understanding of educational service have all transformed since this initial decision. Through multiple collaborations involving international and local partners, Hosman and her teams of students have designed and implemented solar-powered educational solutions that help overcome critical impediments (e.g., scarcity, remoteness, and geographic isolation) facing individuals in small partnering countries. The stories will inspire those asking, 'Where do I start?' and 'What does it take?' to become a practitioner-scholar in the field of education technology for international development.

5 Concluding Remarks

The implementation of ICT internationally can be seen as influenced by delocalizing global dynamics that are themselves made possible through technology. ICT have the potential to connect people by removing the barriers to communication and social interaction. ICT also have powerful enabling effects in delocalizing trade and economic activity through the emergence of transnational markets, including for education. In so doing, the relevance of individual nation states is diminished. However, consider that for small countries, their comparative and competitive advantages reside in their localization: their unique, indigenous cultures, and their local knowledge. Their value rests also in their indigenous human capital that not only must be trained and expanded, but also retained, in order to contribute to development. These contrasting arrangements are the foundations of critical misalignment of the global education agenda with the local development aspirations of small countries. Global dynamics have transformed education systems in small developing countries, calling for more analytical evaluations of the promised improvements in learning outcomes. While it is important to expand international participation in education, issues of local context and culture cannot be sidelined when designing development policy (Crossley, Bray, & Packer, 2009; Heyneman, 2009) – this holds true for ICT in education policies.

The chapters presented in this volume all point to the role of ICT in providing expanded access to education in the developing world and also emphasize the need for education quality, equality, and equity. The authors argue for stepping away from uncritical technology implementations that precipitate new digital divides and reinforce existing inequalities. Rather the authors invite us to consider ICT implementations that have, at their core, a focus on the social arrangements of recognition, respect, and self-directed participation in teaching and learning in developing countries. The economic conditions and other contextual circumstances in small developing states make it difficult to implement, evaluate, replicate, and scale up ICT-supported innovations in education. The chapters herein reveal important lessons that stem from some promising domestic ICT policy, development, and implementation endeavors as well as some mutualizing international partnerships in education. Small countries can teach big lessons that could be missed due to our own inattention. It is hoped that the work in this volume will alert our gaze toward these important issues, places, and people.

AcknowledgementsThe author would like to thank the following individuals for their comments and other assistance concerning this chapter including Luanne Serieux-Lubin, Alvinus Melius, and Michael Thomas. However, the views expressed in this chapter and any shortcomings are those of the author.

References

Augustin, A. (2007). *Globalization challenges for small island developing states* (Unpublished thesis). University of Trier, Germany.

Bacchus, M. K. (2008). The education challenges facing small nation states in the increasingly competitive global economy of the twenty-first century. *Comparative Education*, *44*(2), 127–145.

BBC News. (2016, October 8). *Hurricane Matthew: Haiti south '90% destroyed'*. Retrieved from http://www.bbc.com/news/world-latin-america-37596222

Becker, G. S. (2008). *Human capital*. Retrieved from http://www.econlib.org/library/Enc/HumanCapital.html

Briguglio, L. (1995). Small island developing states and their economic vulnerabilities. *World Development*, *23*(9), 1615–1632.

Central Intelligence Agency (CIA). (2017). *The world factbook*. Available from https://www.cia.gov/library/publications/the-worldfactbook/fields/2012.html#al

Clarke, S., Wylie, G., & Zomer, H. (2013). ICT 4 the MDGs? A perspective on ICT' role in addressing urban poverty in the context of the Millennium Development Goals. *Information Technologies and International Development*, *9*(4), 55–70.

Crossley, M., Bray, M., & Packer, S. (2009). Education in the small states of the Commonwealth: Towards and beyond global goals and targets. *The Round Table*, *98*(6), 731–751.

Crossley, M. W. (2010). Context matters in educational research and international development: Learning from the small states experience. *Prospects*, *40*(4), 421–429.

Crowards, T. (2002). Defining the category of small states. *Journal of International Development*, *14*, 143–179.

Di Battista, A., Dutta, S., Geiger, T., & Lanvin, B. (2015). The Networked Readiness Index 2015: Taking the pulse of the ICT revolution. In S. Dutta, T. Geiger, & B. Lanvin (Eds.), *The global information technology report* (pp. 3–30). Geneva, Switzerland: World Economic Forum.

Dodson, L., Sterling, S. R., & Bennett, J. K. (2012). Considering failure: Eight years of *ITID* research. *Information Technology and International Development*, *9*(2), 19–34.

Easterly, W., & Kraay, A. (2000). Small states, small problems? Income growth, and volatility in small states. *World Development*, *28*(11), 2013–2027.

Everest-Phillips, M. (2014). *Small, so simple? Complexity in small island developing states*. Singapore: UNDP Global Centre for Public Service Excellence.

GSM Association (GSMA). (2008). *The GSMA Development Fund Top 20: Research on the economic and social impact of mobile communications in developing countries*. London, England: Author.

GSM Association (GSMA). (2013). *Scaling mobile for development*. London, England: Author.

Heyneman, S. (2009). The failure of education for all as political strategy. *Prospects*, *39*, 5), 5–5),10.

Higgins, S., Xiao, Z., & Katsipataki, M. (2012). *The impact of digital technology on learning: A summary for the Education Endowment Foundation*. London, England: EEF.

Independent Evaluation Group (IEG). (2011). *Capturing technology for development: An evaluation of World Bank group activities in information and communication technologies, The evaluation* (Vol. 1). Washington, DC: Author.

International Monetary Fund (IMF). (2015). *World economic outlook*. Washington DC: Author.

Jansen, M. (2004). *Income volatility in small and developing economies: Export concentration matters*. Geneva, Switzerland: World Trade Organization.

Kirkman, G. S. (2002). *Networked readiness and small island developing states*. Paper presented at the Conference on Small Island Economies in the Era of Globalization, Center for International Development at Harvard University, Cambridge, MA.

Kozma, R. (2005). National policies that connect ICT-based education reform to economic and social development. *Human Technology*, *1*(2), 117–156.

Kozma, R., & Wagner, D. (2005). Core indicators for monitoring and evaluation studies for ICT in education. In D. Wagner (Ed.), *Monitoring and evaluation of ICT in education projects – A handbook for developing countries* (pp. 35–57). Washington, DC: Infodev.

Lowther, D. L., Inan, F. A., Strahl, J. D., & Ross, S. M. (2008). Does technology integration "work" when key barriers are removed? *Educational Media International, 45*(3), 195–206.

Lowther, D. L., Ross, S. M., & Morrison, G. R. (2003). When each one has one: The influences on teaching strategies and student achievement of using laptops in the classroom. *Educational Technology Research and Development, 51*(3), 23–44.

Lubin, I. A. (2016). *Intentional ICT: Curriculum, education and development* [IBE Working Papers on Curriculum Issues No. 17]. Geneva, Switzerland: IBE-UNESCO.

Nottage, H. (2014). *Commonwealth small states and least developed countries in the WTO dispute settlement system* [Commonwealth trade policy discussion paper 2014/02]. London, England: Commonwealth Secretariat.

Organisation for Economic Co-operation and Development (OECD). (2015). *Frascati Manual 2015: Guidelines for collecting and reporting data on research and experimental development.* Paris, France: Author.

Organization of American States, Executive Secretariat for Integral Development (OAS-SEDI) (1997). Small and relatively lesser developed economies and western hemisphere integration. Retrieved from http://www.sedi.oas.org/DTTC/TRADE/PUB/STUDIES/SECON/INDEX.asp

Rangaswamy, N., & Cutrell, E. (2012). Anthropology, development, and ICTs: slums, youth, and the mobile Internet in urban India. *Information Technology and International Development, 9*(2), 51–63.

Shileds, R. (2013). *Globalization and international education.* London, England: Bloomsbury.

Tamim, R. M., Bernard, R. M., Borokhovski, E., Abrami, P. C., & Schmid, R. F. (2011). What forty years of research says about the impact of technology on learning: A second-order meta-analysis and validation study. *Review of Educational Research, 81*(3), 4–28.

The Guardian. (2017). *How the Caribbean islands are coping after hurricanes Irma and Maria.* Retrieved from https://www.theguardian.com/world/2017/sep/21/caribbean-islands-hurricane-irma-maria-puerto-rico

Umemura, T. (2016). Small island developing states and globalization: Development potential. In M. Ishihari, E. Hoshino, & Y. Fujita (Eds.), *Self-determinable development of small islands* (pp. 133–158). Singapore: Springer.

UNESCO Institute for Statistics (UIS). (2017). *Literacy rates* [Extracted data]. Available from http://data.uis.unesco.org/Index.aspx?queryid=166

United Nations Conference on Trade and Development (UNCTAD). (2014). *Trade and development report, 2014.* New York, NY/Geneva, Switzerland: Author.

United Nations Department of Economic and Social Affairs (UNESA). (2015). *World population prospects, the 2015 revision.* Available from https://esa.un.org/unpd/wpp/DataQuery/

United Nations Economic Commission for Latin America and the Caribbean (ECLAC). (2015). *Report of the symposium on sustainable development goals for the Caribbean within the post-2015 development agenda.* Port-of-Spain, Trinidad & Tobago: Author.

United Nations Educational, Scientific and Cultural Organization (UNESCO). (1990). *World declaration on education for all and framework for action* (Adopted by the World Conference on Education for All, Jomtien, Thailand). Paris, France: Author.

United Nations Educational, Scientific and Cultural Organization (UNESCO). (2000). *Dakar framework for action. Education for all: Meeting our collective commitments* (Adopted by the World Conference on Education for All, Dakar, Senegal). Paris: Author.

United Nations Office of the High Representative for the Least Developed Countries, Landlocked Countries and Small Island Developing States (UN-OHRLLS). (2011). *Small island developing states: Small island big(ger) stakes.* New York, NY: Author.

United Nations Office of the High Representative for the Least Developed Countries, Landlocked Countries and Small Island Developing States (UN-OHRLLS) (2014). *The vulnerability of Landlocked developing countries to external shocks.* Retrieved from http://www.lldc2conference.org/custom-content/uploads/2014/10/VulnerabilityReport.pdf

United Nations Office of the High Representative for the Least Developed Countries, Landlocked Countries and Small Island Developing States (UN-OHRLLS) (2016). *Releasing trade potential in landlocked developing countries vital to future sustainable development* [Press Release]. Retrieved from http://unohrlls.org/news/22-june-2016-ministerial-meeting-landlocked-developing-countries/
United Nations Office of the High Representative for the Least Developed Countries, Landlocked Countries and Small Island Developing States (UN-OHRLLS) (2017a). *Country profiles*. Available from http://unohrlls.org/about-lldcs/country-profiles/
United Nations Office of the High Representative for the Least Developed Countries, Landlocked Countries and Small Island Developing States (UN-OHRLLS) (2017b). *About the small island developing states*. Retrieved from http://unohrlls.org/about-sids/
West, M., & Chew, H. E. (2014). *Reading in the mobile era: A study of mobile reading in developing countries*. Paris, France: UNESCO.
Winthrop, R., & Smith, M. S. (2012). *A new face of education. Bringing technology into the classroom in the developing world* [Working paper]. Washington, DC: Brookings Institution.
World Bank. (2017a). *Country and lending groups* [Country classifications]. Available from https://datahelpdesk.worldbank.org/knowledgebase/articles/906519#High_income
World Bank. (2017b). *Data indicators*. Available from http://data.worldbank.org/indicator
World Bank Group. (2012). *ICT for greater development impact: World Bank Group strategy for information and communication technology: 2012–2015*. Retrieved from http://documents.worldbank.org/curated/en/285841468337139224/pdf/715400WP0WBG0I0sclosed0July02502012.pdf
World Bank/Commonwealth Secretariat (2000). *Small states: Meeting challenges in the global economy*. Report of the Commonwealth Secretariat/World Bank Join Task Force on Small States. Retrieved from http://documents.worldbank.org/curated/en/267231468763824990/pdf/270290Small0St1also020318001public1.pdf
World Economic Forum. (2015). *The global information technology report 2015. ICTs for inclusive growth*. Retrieved from http://www3.weforum.org/docs/WEF_Global_IT_Report_2015.pdf
World Trade Organization (WTO). (2002). *Small economies: A literature review* [Dedicated session WT/COMTD/SE/W/4, document WT/L/447]. Geneva, Switzerland: Committee on Trade and Development, WTO.

Author Information

Ian A. Lubin
Complete mailing address: Independent Scholar and Researcher
1305 Barnard St. # 81, Savannah, GA
31401, USA
Telephone number: 912-386-1047
Email address: ianlubin@gmail.com

Ian A. Lubin is a professional educator and research scientist specializing in learning and instructional psychology, educational technologies, and international education development. He has served as an instructional faculty and researcher at various national and international institutions. Dr. Lubin is the author of several journal articles, chapters, working papers, and other consultation publications on the design of optimal learning environments, cognition and motivation for learning, utilization of instructional technologies, teacher/faculty professional development, and education for development in international contexts. Dr. Lubin serves as a specialist advisor to international development organizations and agencies such as UNESCO and the Commonwealth of Learning. He holds undergraduate degrees in Psychology and Multimedia Instructional Design and earned his M.Ed. and Ph.D. degrees in Instructional Psychology and Technology.

Chapter 2
The Research Agenda for Technology, Education, and Development: Taking Stock and Looking Ahead

Francesc Pedró

Abstract This chapter discusses the role that research should play in the current discussions about the role of technology in educational development, with particular reference to small developing states. It shows that often decisions about technology use in education are far from being based on evidence and suggests that this is due partly to the absence of a rigorous knowledge base but also of an appropriate research agenda. No doubt, research has a role to play investigating further the role that technology solutions play in improving the quality of education. But the right research questions must be asked. This chapter addresses two questions. First, it looks at what is currently known and the limitations of the existing knowledge base about how technology can contribute to improving learning – recognizing the paradox that developing countries, which make comparatively bigger efforts in this domain, lag also in terms of knowledge base. The question is therefore whether research has missed the point and failed to build a comprehensive, coherent and useful knowledge base. Second, the chapter considers both what elements are missing and how the important research challenges required to assemble them could be met.

Keywords Educational technology • Developing countries • Educational research

1 Introduction

Technology's potential to transform education has been declared numerous times, but it is widely accepted that for various reasons this potential has not yet been harnessed as expected, neither in the developed countries (Conlon & Simpson, 2003; Cuban, 2001; OECD, 2015; Sandholtz, 2001) nor in developing countries (Power, Gater, Grant, & Winters, 2014). When reviewing how decisions about technology

F. Pedró (✉)
UNESCO, 7 place de Fontenoy, 75007 Paris, France
e-mail: f.pedro@unesco.org; http://en.unesco.org/

I.A. Lubin (ed.), *ICT-Supported Innovations in Small Countries and Developing Regions*, Educational Communications and Technology: Issues and Innovations,
DOI 10.1007/978-3-319-67657-9_2

use in education are made, it is striking how little is known about the effects of using technology[1] on the quality of school education and,[2] more specifically, which particular uses of technology can result in better learning. Not only policy makers but also teachers and parents alike are left with the idea that there is no conclusive evidence or, in the best case scenario, no significant difference between using and not using technology, which some take as a clear assertion that at least technology does not cause any harm to students. Especially related to small states and developing regions, if a good evidence-supported knowledge base existed in this domain, then the analysis of these effects and the factors that determine or condition them could be used to unveil what works and why. But in the absence of hard evidence the evaluation of these policies, no matter if they are at national, school, or classroom levels, remains an almost impossible endeavor and the whole issue of how policy decisions are made remains open. This is a far from optimal state of affairs in poorer, resource-constrained developing contexts, where technology-based reforms are being pushed as the remedy for poor economic and social conditions. As the emphasis of many national initiatives in these contexts is usually put on granting access to technology, as an intrinsic added value, not much research on the actual effects on learning has been conducted.

This raises the question of how to synthesize the available evidence on how technology is used to improve learning while revealing the gaps. One way to respond is to analyze the issue on three possible levels of research: the most basic is to look at state-of-the-art reviews; beyond that level, there are the meta-analyses, all characterized by their sole reliance on empirical research that fulfills certain basic requirements of methodological rigor (which dramatically reduces the number of studies considered); and finally, there are second-order meta-analyses, that is to say syntheses of meta-analyses.

1.1 The Existing Knowledge Base

Research on the relationship between technology and education is not in its infancy; on the contrary, countless research projects have been dedicated to this question. As has been pointed out quite often, the existing knowledge base is scattered and limited in scope and the general feeling is often that conclusive evidence of the success of technology on learning outcomes has not yet been discovered (Falck, Mang, & Woessmann, 2015; OECD, 2015). Indeed, there have not yet been many significant internationally recognized attempts to summarize the knowledge gained from

[1] References to technology are to be understood as limited to digital equipment, services, contents, and applications to support teaching and learning.

[2] This contribution will refer solely to basic schooling (typically, compulsory primary and secondary education), considering that the characteristics of higher education and other levels and forms of formal and non-formal education, such as technical and vocational education or adult education, for instance, would require a very different approach.

empirical evidence to answer the simple question of what we have learned from the decades of research on the relationship between technology and education, particularly in developing countries.

1.2 State-of-the-Art Reviews

Most attempts to summarize the state of the art have tried to answer the question of how research has progressed on very specific themes, such as comparing learning results achieved through traditional methods and those assisted by technology (Bernard, Abrami, Lou, & Borokhovski, 2004; Bernard et al., 2009; Tamim, Bernard, Borokhovski, Abrami, & Schmid, 2011); technology-assisted teaching (Bernard et al., 2009), or simply reviewing results achieved by specific technological solutions (Dillon & Gabbard, 1998; Fabos & Young, 1999) as well as analyses of the effects of using these solutions in ongoing teacher training (Lawless & Pellegrino, 2007; van Rooij, 2009).

Unfortunately, the majority of these state-of-the-art reviews tend to approach the question more from a descriptive perspective (who researched what and how they did it), rather than an analytical one and, therefore, do not provide a synthesis of available evidence. Among all these synthesis attempts, the most recent is that by Hsu et al. (2012). The synthesis was limited to what are probably the five most significant indexed international publications in this area, solely during the period from 2000 to 2009; however, the analysis of their content is sufficiently illustrative. On the one hand, the results demonstrate that in this field less attention has been paid to empirical research on impacts than to the theoretical study of models and architectures, which may be seen as evidence of the intentionally exploratory and non-experimental nature of the majority of research in this area. On the other hand, the overwhelming majority of the research has focused on higher education. Similar summary conclusions and reviews, although limited to more specific areas such as mobile learning (m-learning) (Hwang & Tsai, 2011), collaborative learning (Keser, Uzunboylu, & Ozdamli, 2011) or, indeed, educational events broadcast via technology (Wu et al., 2013), lead to practically identical conclusions.

The most interesting aspect, in terms of a concordant characteristic of all these reviews, is that they attempt no synthesis of the knowledge base cumulated through research. In other words, these reviews fail to answer the question of what we know as a result of the accumulated evidence gathered, as is done in other areas of research. Rather, they merely describe the themes that have been the objects of research, without really drawing conclusions or lessons learned. Strangely, these attempts, in light of the fragmented nature of the research, ultimately conclude that what is needed is ... more research. This is, for instance, what a 2015 working paper on digital education concludes: “Overall, scholars highlight a need for further research to evaluate the effectiveness of digital learning on K-12 student outcomes” (Center for Promise, 2015, p. 7).

1.3 *First and Second-Order Meta-analyses*

Beyond the most basic level of synthesis that can be found in these state-of-the-art reviews, there is a lack of meta-analysis, at least in the strict sense of the term (Lipsey & Wilson, 2001), especially second-order meta-analysis. The most well-known meta-analysis in this area is Clark's (1983), which is considered to be the first successful attempt to summarize the impact of the integration of technology on students' results, based on empirical evidence. Curiously, his conclusions seem to be aging well and still completely valid. Many years later came the meta-analysis of Schacter and Fagnano (1999), who carried out a qualitative review of the meta-analyses available at that time. More recently, the most well-known and often-quoted analysis appeared, that of Hattie (2009), which includes some aspects related to technology and learning in this synthesis of the meta-analyses of education research. Hattie examined 76 meta-analyses of 4498 different studies, involving a total of 3,990,028 students and teachers from studies carried out over a period of 30 years. Using effect sizes as an indicator, he was able to identify six clear areas where the use of technological components seemed to lead to improvements in students' learning. In short, Hattie concluded that technology has a positive effect on learning when the following conditions are met:

1. A variety of teaching strategies is used. If students are dealing with at least two different strategies, they will practice a *deliberative approach*. Technology, in this sense, does not replace the teacher but rather reinforces his or her presence.
2. The teacher has received initial training in the use of technology, not only as a tool for processing information and communication but also as a teaching and learning tool.
3. Multiple resources for learning, such as tutorials, practical sessions, role-play, and problem-solving are available.
4. The student, not the teacher, controls the timing, rhythm, and order of the lesson.
5. Learning is optimized between peers, for example, using mixed groups. In any case, learning supported by technological resources is most effective when not done in isolation.
6. The use of feedback is optimized, so that the student receives explanations and alternative corrections, instead of just being given the expected correct answer.

Still, there have been few other attempts to compile what is known in order to construct an easily accessible synthesis. The most recent is the meta-analysis carried out by the team at Edutopia. In this analysis a few general principles on what seems to be accredited in different areas of school education are defined (language, mathematics, sciences, digital skills), resulting in a complex overview. The two basic principles emerging from this review are very telling:

- The mere incorporation of technology into classrooms does nothing to improve learning. Rather, what is important is how both students and teachers use technology to develop knowledge and skills. The satisfactory integration of

technology in learning, in general, goes hand in hand with teacher training, study plans, methodological strategies, and the assessment of learning (Bebell & O'Dwyer, 2010; Zucker & Light, 2009).
- In a school setting, the use of technological resources that enrich teaching and learning practices gives better results than the replacement of the physical presence of the teacher by a technological resource (Cheung & Slavin, 2011; Cheung & Slavin, 2013; Means, Toyama, Murphy, Bakia, & Jones, 2009).

Second-order meta-analyses are simply not as numerous as one might expect given the high number of available academic journals on technology and education. In fact, there is only one that fulfills all the classic requirements of a second-order meta-analysis: Tamim et al. (2011). This important work, which like Hattie's considers effect sizes, draws three main conclusions. Firstly, in trial conditions, the groups with which technology was used achieved significantly better learning results, although the impact was small or, in many cases, moderate but never clearly significant. Secondly, the use of technology to support teaching leads to significantly better results than the use of technology for direct instruction, that is to say, as the only means of instruction, replacing teachers. Lastly, these positive effects seem to be greater in primary than in secondary schools.

1.4 Learning from What Works in Developing Countries

Initiatives centered on the mass distribution of equipment are ever more prevalent in developing countries, with rather disappointing results (Arias Ortiz & Cristia, 2014; Cristia, Ibarrarán, Cueto, Santiago, & Severín, 2012; Winthrop & Smith, 2012). This is probably due to their symbolic value, which is greatly appreciated by politicians: they give the public a clear message of intent to modernize education, which is easily translated into figures, through the delivery of devices, creating visible achievements. However, the accompanying evaluations have not yet provided conclusive data. They probably never will, for the simple reason that the determining factors of impacts on learning are the quality of instructional designs and the capacities of the teachers that put them into practice – and these are not considered as variables either in assessments or, unfortunately, in research. Owing to increased investment in technology and the fact that it has not always been successful, the idea has gradually emerged that more effort is needed to measure and evaluate technology's impact. Although this is undeniably the case, in reality, the strategies used so far continue to be methodologically weak (Guislain, Zhen-Wei Qiang, Lanvin, Minges, & Swanson, 2006) to the point that appropriate plans are not always made for the incorporation of baseline studies that define a starting point and then allow for the identification of progress made, monitoring mechanisms, and a rigorous evaluation of the impacts.

The analysis of the effectiveness of various interventions for improving student learning in developing countries has been the object of a myriad of studies that has been reviewed several times in search of coherence (Conn, 2014; Ganimian & Murnane, 2016; Glewwe, Hanushek, Humpage, & Ravina, 2013; Kremer, Brannen, & Glennerster, 2013; Krishnaratne, White, & Carpenter, 2013; McEwan, 2015). Quite recently Evans and Popova (2015) have documented and reviewed 301 studies from developing countries: 227 of those studies reported learning outcomes and 152 reported enrollment or attendance outcomes. In most of these studies, the provision and use of technology serving different needs and purposes are featured among other possible interventions. The researchers conclude that there is ample support across the reviews for the success of (i) pedagogical interventions, including through the use of technology; (ii) individualized teacher training; and (iii) accountability-boosting interventions.

Despite this support, the research focused heavily on regression analyses, and it is difficult to elucidate the significance of technology use simply by non-experimental data (Glewwe, 2002). Importantly, the conclusions reached in these studies are in no way different from the ones dealing with developed countries, leading once again to the *non-significance phenomenon* situation. Therefore, it is worth looking at randomized and controlled experiments as a better research strategy to understand what difference the use of technology can make in the classrooms of developing countries.

In a meta-analysis of randomized experiments related to primary schools in developing countries (McEwan, 2015), the mean effect size for technology-related interventions was found to be the highest (0.15) among all the reviewed instructional interventions, including, in addition to technology use, teacher training, textbooks, and/or changes in the size and composition of classes. Yet, two observations have to be made: first, that the number of experiments reviewed was quite low – only 13 until 2014 complied with the methodological requirement of being randomized, controlled experiments[3]; and second, that the geographic reach was also quite limited, with an ample majority of the reviewed experiments located in China and India, where the most encouraging results are found (Banerjee, Cole, Duflo, & Linden, 2007; Lai et al., 2012; Linden, 2008). Unfortunately, there is a lack of comparable research on small developing countries. One major reason for this lies in the costs of carrying out often complex monitoring and evaluation procedures when the emphasis of the initiatives is mostly to grant access to technology in schools, rather than to improve learning outcomes or to promote pedagogical change.

Clearly, instructional materials alone do not improve learning, neither does technology (Glewwe, Kremer, & Moulin, 2009; Kremer et al., 2013). Yet they appear to be effective only when supported by a well-articulated instructional model and the corresponding teacher-training program (Banerjee et al., 2007; Friedman, Gerard, & Ralaingita, 2010; Lucas, McEwan, Ngware, & Oketch, 2014) that critically affects student's learning experience. In line with this, Ganimian and Murnane (2016) in a review of rigorous impact evaluations also concluded that more or better

[3] Typically, a random assignment of students or schools to school-based interventions improves the internal validity of causal inferences (Duflo, Glennerster, & Kremer, 2008; Glewwe & Kremer, 2006).

resources, including technology, improve student achievement only if they result in changes in children's daily experiences at school – a real challenge for the majority of schools in developing countries. To improve student achievement, resources such as technology, textbooks, libraries, or flipcharts first have to be used, which is not always the case, and second, they should change children's daily experiences at school, i.e. change students' learning experiences and be adapted to individual student learning levels (Evans & Popova, 2015).

1.5 Summing Up: A Weak Knowledge Base that Does Not Factor in Pedagogy

In conclusion, the resulting picture we are left with by this synthesis of reviews is very nuanced. It seems that Reeves' assertions from the turn of the century (Reeves, 2000) still hold true in terms of the three fundamental issues in research on the relationship between technology and education. Firstly, there is great confusion regarding the differences between basic and applied research. Secondly, the quality of research in the field of educational technology is generally poor (although no poorer than education research in other areas). Thirdly, syntheses on educational technology research, as with literature reviews and meta-analyses, only provide professionals with an insufficient outlook or one which ends up being confusing or inconclusive. The knowledge based on evidence provided by research is scarce and dispersed (Cradler, 2003). For example, there are currently very few documented cases of a systemic increase in students' learning that can be directly attributed to technological innovations. In more than one sense, this is probably another example of inflated if not incorrect expectations, supported by the growing investment in facilitating access to equipment, initially, and then to connectivity, as well as messianic hopes and idealist reforms (Pedró, 2012).

The overall impression, which appears to be common sense but is supported by numerous evaluations (Kozma & Wagner, 2005), is that there is no consistent relationship between the mere availability of technology and an improvement in learning results. However, beyond this generic assertion, there are some very important gaps in our knowledge of what works and what does not, and why. In fact, the question of impact is by its very nature difficult to measure and permanently open to debate. If the question relates to the supposed superiority of technology-supported teaching – in short, the question 'is it better?' – the answer is 'only sometimes, depending on other factors'. The issue is whether the right balance is struck between the different key components: usually between the provision of hardware and software on the one hand and educational capacity and support on the other.

Teachers' roles are key to transform more resources into better learning opportunities. Yet they may feel lost, more so in small developing countries, if they are not adequately guided and supported to make the best use of those resources. This is exactly what Berlinski, Busso, and Cristia (2013) found when comparing the effects of guided versus non-guided programs of technology-based interventions in developing countries: programs that guide technology use produce substantially greater increases in learning than those focused merely on providing resources, whose use is left at the discretion of the school or, more often, the individual teacher. Moreover, McEwan (2015) provided empirical evidence that guided-use programs are among the educational interventions with the highest impacts on student achievement, while nonguided-use programs hardly produce any positive effect.

What lies behind this never-ending discussion is a revisited version of the productivity paradox. The classical productivity paradox was formulated to describe the phenomenon of the lack of relationship between the adoption of a new technology and the productivity gains in a business or industry context. As Robert Solow (1987, para. 7) put it, "you can see the computer age everywhere but in the productivity statistics." A number of studies carried out in the 1990s found it difficult, or otherwise impossible, to discern, in aggregate economic performance data, the productivity gains that were expected after massive investments in technology (Brynjolfsson, 1993; Landauer, 1995; Triplett, 1999). The paradox was explained in part by the fact that the investment took some time to be fully applied and exploited because there were lags in investment in the skills of workers, the structure of work and organizations, and business strategies. It was thus concluded that much of the payoff of investments in technology would come, not in performing the same functions more efficiently, but by performing new functions that often reflected transformations in processes.

In short, what is expected to follow the adoption of a new technology, and what justifies the corresponding investment, is the expectation of seeing an increased productivity. But what actually happens is that if a new technology is adopted in a context where processes are kept unchanged, technology may be found to be useless, if not obtrusive, and in many cases, this may even lead to an actual decrease in productivity.

When it comes to education, the productivity paradox becomes the educational productivity paradox (Peslak, 2005) or the student productivity paradox (Hikmet, Taylor, & Davis, 2008). The point is that, in education, technology is a blind tool or artifact that can be used for a variety of purposes. Whether the adoption of technology is linked or not to educational performance will depend on the improvements associated with changes in methodology, which would require appropriate technical and pedagogical support. If the methodology remains the same as it was before the introduction of technology, as is often the case when teachers adopt technology in order to perfect what they were already doing, expectations are low. If in addition to this, the intensity of use of technology is reduced, the amount of preparation time and effort may not be compensated by the educational benefits obtained.

It is important to insist on the critical importance of the instructional uses given to technology as the key factor in explaining what are the effects of technology use on student achievement. For instance, in line with most of the literature, Falck et al. (2015) on average do not find a significant effect of technology use in the classroom on student achievement. They hypothesize that this null effect is the combination of positive and negative effects of specific uses – in short, it is the pedagogy that matters. If the time spent on technology use is increased, it substitutes different alternative time uses. There is, therefore, a time-opportunity cost and, thus, the net effect of technology use depends on the specific activities that it is used for and the relative instructional effectiveness of the activities that they allow to perform. An overall null effect of technology use – yet another instance of the non-significance phenomenon – may be the sum of positive and negative effects.

Their results (Falck et al., 2015) can help reconcile some of the contending findings in the literature, that either find little to no effect of technology use on student achievement or rather a negative influence. There are however exceptions: studies finding significant positive effects of specific guided uses that focus on effective activities. Moreover, the result that the effects of technology use in the classroom differ by their specific nature are also in line with the recent literature on technology use at home, which emphasizes that home devices can be put to conducive uses such as schoolwork as well as to alternative uses such as gaming or entertainment that yield no positive results (Faber, Sanchis-Guarner, & Weinhardt, 2015; Fairlie & London, 2012; Fairlie & Robinson, 2013; OECD, 2010a).

At this point, it is worthwhile to return to the recognized and pioneering work of Clark in the early 1980s (1983). He conducted a meta-analysis on research in educational technology and reached the conclusion that technology was a "mere vehicle that delivers instruction but does not influence student achievement any more than the truck that delivers our groceries causes changes in our nutrition" (p. 2). This *transmissive* perspective considers technology to be a delivery mechanism without the ability to intrinsically influence the quality of learning.

This point of view is in sharp contrast with the vision that has traditionally guided the majority of research in this field, which could be considered, as an *exposure model* for educational technology in that it suggests that exposure to technology would affect learning. In the end, however, Clark's conclusion is common sense: mere exposure to technology has no educational benefits. Of course, the same can be said of pencils, paper, books, chalkboards, and all other technological devices that are typically used in schools. We cannot expect that simply placing a book and a student in the same room will necessarily result in learning, although obviously, this could happen by chance. Exposure to books is a necessary condition, but not sufficient, to make books educationally effective (Global Education Monitoring Report, 2016). There are clearly other critical variables such as the content of the book or the way in which the child interacts with this content, as well as the guidance provided by the teacher to help the student benefit from it. The same can be said for technology. To use Clark's analogy, for a truck delivering groceries to be effective in improving the nutrition of a given person, the person must live on the truck's delivery route and it must carry something other than doughnuts, hamburgers, and chips.

Clark's observation leads to a sweeping conclusion regarding research: considered in isolation, technology probably has no generic effect on teaching and learning. For impacts to be observed, at least two other factors must be taken into consideration: the content and the instructional strategies used. Thus, research questions and models that do not differentiate between the content being taught, the instructional strategies used, and the way in which technology interacts with these variables will most likely continue to show that the mere use of technology does not lead to educational benefits, or will at least give greater weight to the conclusion that there is *no significant difference* (Russell, 1999), as is so often repeated in the field of educational technology. Only those studies which explore how technology interacts with teaching and content can begin to challenge Clark's assertion that "media [...] do not directly influence learning" (Clark, 1983, p. 2), by making clear that the uses of these media can be significant. Take the following example: in a study on the impact of the use of video streaming on students' learning (Boster, Meyer, Roberto, Inge, & Strom, 2006), it was concluded that there was no significant difference between the control and the experimental group. The study, however, gave no information on the educational content or the pedagogical use of technology, which makes the study impossible to replicate. It is also most likely that different teachers used different instructional strategies. In conclusion, when the instructional methods used are not specified, it becomes difficult to understand the implications or determine how the results could be reliably replicated in other contexts. This example shows that poorly oriented research, but with an experimental design, can be published thus clouding those who are looking for guidance.

The right question, therefore, is not which new technology leads to increased productivity, but which new technology-supported methodologies improve student performance over traditional ones if any at all, and which other factors intervene. Previous calls have already been made in order to investigate the explicit relationships among technology, instructional strategy, psychological processes, and contextual factors – as for instance by Alavi and Leidner (2001). The almost infinite array of methodological possibilities makes this kind of investigation extremely difficult but not impossible, provided that there is sufficient effort devoted to the accumulation and dissemination of the resulting knowledge base. This is a major task not only for educational research but also for teacher training institutions, which should contribute to the dissemination of the results obtained. Such a task might appear overwhelming, particularly as the technological frontier is constantly changing. However, it is worth the effort.

2 What Should Be the Fundamental Research Question?

There is one classical research question in this domain: does technology-supported education make a difference? Or, more generally, does technology lead to better student results? The investigation and elucidation of comparative advantages is in fact at the core of many scientific studies; a simple question such as 'what is better?' is naturally very attractive to political decision-makers and inexperienced researchers (Schrum et al., 2007), but the answer is not as simple as the question seems to be. As seen in the previous section, when looking for a response in the existing knowledge base about the effects of technology in education, a striking fact seems to emerge: there is no conclusive evidence. This has been known for some years as the 'non-significance phenomenon,'[4] leading to the overall conclusion that, in education, technology makes no difference because the investments made have not translated into improved educational productivity (Brynjolfsson, 1993; Hikmet et al., 2008; Peslak, 2005), thus reasserting Solow's productivity paradox in the education sector (Triplett, 1999).

Although it appears that the answer to such a question (whether technology improves students' results or not) should be quite straightforward, the problem is that the question is neither logical nor useful because it is not formulated in a way that takes into account the complexity of education. Additionally, and more importantly, it does not help to inform policy decisions in an appropriate way, leaving unanswered the issue of whether the investments in technology are worth the effort. Briefly, when phrased this way, without caution, the issue leads to confusion because it is ill-defined or, at least, not sufficiently defined for a proper empirical assessment.

Behind all this confusion is a poor understanding between those who make the decisions about investments and those who are expected to benefit from them, because two different rationales are quite often in place. The former would like to receive a clear-cut response about whether or not the investments in educational technology pay off. For the decision makers, these investments are one possible option among many. Teacher training, career incentives, salary increases, or a decrease in class size are just some of the alternative options for educational investments that policy makers can use to improve the quality of education. To make informed decisions, policy makers need cost-effectiveness analysis – although their final decisions could also take into account other factors, such as the symbolic value of the proposed policy. For them, the issue is very straightforward: is it worth investing in educational technology?

Teachers and educationalists look at the issue very differently. They usually insist on the complexity of education – that is, on the fact that learning is not only a function of formal teaching activities that take place in the classroom but also of other nonformal educational influences that can hardly be accounted for in the productivity equation of school education. Activities that are undertaken by the learner outside school – and even inside school – at home or with peers, interact with those carried out during formal instruction hours. As a result, when learning results are measured – as in the context of national or international student assessments – it is almost impossible to isolate the role played by one individual teacher in a specific subject matter during one particular academic year.[5] For the same reason, it is also impossible to isolate the effects of the technology components in learning from the most important factor: the strategies put in place (Falck et al., 2015).[6]

[4] This expression was first used to indicate that, in the particular case of distance education, research was unable to demonstrate the superiority of technology-supported courses over traditional arrangements of distance education. It has been later extended to all other areas of education (Russell, 1999).

[5] However, different techniques related to the measurement of added value aim at improving the chances of isolating influences.

[6] Technology components include both traditional (such as paper and pencil) and digital.

All this becomes even more complex in the context of developing countries. The challenges of adopting technology cannot be overemphasized when discussing schools in poor and geographically remote places, especially in small developing regions. There is a general lack of resources, training for teachers, connectivity, electricity, and the list goes on. The challenges are not just for technology but also for other aspects. Some of these places do not really have a history of technology use to draw from. There is a clear challenge in research as well, as the difficulties of assessing the strategies that may claim to work in these developing classrooms have to confront, again, the issue of the impossibility of isolating the technology factor from the rest of intervening elements. But even when this can be done, the fundamental question involves whether a technology push in education is really what schools need as opposed to other alternative investments (such as more and better textbooks or a better-trained teacher force), or not. When entering this debate, one has to bear in mind that, in particular in small developing states, the scenario of not connecting schools is far worse than having connected schools with little or no improvements in learning outcomes. As some examples of small states show, the deployment of technology in schools may serve well other non-instructional goals, such as equity in access to an asset that many see already as a human right.

Therefore, the issue about the effects of technology in education has to be rephrased in a way that takes into account the complexity of the intervening factors in learning and the overall consideration that access to technology represents, at least in the context of small developing states, more than an add-in to a plethora of instructional resources. The real question is not about whether to use technology or not, but about teaching and learning strategies and the ways in which technology solutions can make them more efficient. In short, the adoption of technology should be instrumental in improving instruction so that students can learn more and better – and this has to be proven empirically. The problem is that no data are available to address this question comprehensively as of yet.

2.1 Advancing a Comprehensive Approach to Examine Technology Effects in Education

To establish a comprehensive approach to examining the effects of technology in education, two important elements are missing. The first missing factor is a sound knowledge about how technology is actually being used in classrooms, which in turn points to the need for a better understanding of the two key variables of *intensity* and *variety* of technology uses. The second missing element is an exploration of the learning outcomes that, going beyond traditional subject areas, are more closely related to technology; these are widely known as twenty-first-century skills. These are quite often seen as the million-dollar questions: which technology policies contribute most to the quality of education? Which uses of technology boost student performance in the different subject domains? Given the endless array of possible combinations of teaching and learning strategies, the only way to operationalize this question is to look at the relationships between variety and intensity of technology uses, on the one hand, and variations in outcomes, on the other.[7]

To know more about the contribution of technology to the improvement of learning results, it is necessary to expand the knowledge base about teaching and learning at large – that is, to learn about what is going on inside the classroom. A better understanding of the strategies that are currently used in the classroom would be not only informative about the intensity and the variety of uses of technology, but would also be crucial for understanding what works in light of student outcomes and determining the added value, if any, that educational technology brings to teaching and learning, both in the classroom and outside it. Therefore, going further than the mere issue of access, there are clearly two areas for which more information is needed: the intensity and the variety of educational technology usage in the classroom. The intensity of use refers to the amount of time and the sequence of use – that is, how much a particular technology solution is being used, either by the teacher or by individual students. The variety of use, in turn, is related to the vast array of uses that a particular technology device or service allows – again, either in the hands of the teacher or the student.

An exploration of the outcomes also requires addressing the growing number of calls for a wider approach to skills formation for the knowledge economy and society. These calls insist on the need to include those twenty-first-century skills in national and international student assessments. Although an international consensus about what these skills are or how to assess them adequately is not yet in place (Ananiadou & Claro, 2010), what emerges from the ongoing discussions is that many but not all of these new skills can be duly trained and assessed only through an intensive use of technology solutions. This need provides an opportunity to assess the actual benefits to be obtained by using technology in education, provided that the emphasis lies on instruction and not on technology.

Confronted with the educational productivity paradox, there is no doubt that a winning methodological strategy to analyze the effects of technology in education will have to be a mixed-methods approach. Although questionnaires addressed to ministries can work relatively well in centralized systems to investigate the scale of the investments or the state of access, they hardly can be of any use when a more comprehensive perspective, including processes and outcomes, is envisaged or when technology policies in education are decentralized – particularly if the competent authority is each individual school, as increasingly seems to be the case. Empirical experiments, large-scale surveys, direct observations, and interviews can be used to address different research questions and, if appropriately combined, result in a holistic approach.

Empirical experiments serve the purpose of investigating which existing technology solutions result in better learning outcomes while keeping other intervening factors constant. Yet they pose the challenge of how to reach a critical mass of evidence from which clear messages can emerge, which is not yet possible

[7] This process assumes that all other variables remained constant.

(Olofsson, Lindberg, Fransson, & Hauge, 2011). As has been repeatedly shown, different attempts to conduct meta-analysis provide inconclusive, if not confusing, messages. This is probably because existing experiments have taken a piecemeal approach, using alternative theories that can hardly be combined into one theoretical body and focusing on small scale and very specific interventions (Ross, Morrison, & Lowther, 2010). National and international efforts should be made to reach scientific consensus about the lessons learned and the way forward.

Large-scale surveys, both national and international, can provide basic data about the intensity and variety of uses of technology, although these also have important drawbacks. The most important problem with these surveys is that the perspectives of teachers and students have to be combined to obtain a balanced picture. Teachers, on the one hand, may be tempted to provide socially appropriate responses that suit the taste of the public authority on which they depend, possibly overstating the use of technology. Students, on the other hand, may not be qualified enough – because of their age and their lack of wide experience and mature judgment and expression – to properly qualify either the intensity or the variety of uses: hence the problems with surveys of primary school children. However, surveys have an important role to play in determining how and for what purpose technology is being used at the system level as well as the internal variance present across territories, schools, or subject areas. If instead of questionnaires dedicated to technology usage, student outcomes assessment surveys were used, then the resulting knowledge base could be boosted by linking outcomes to intervening factors and thus set the foundations for a general theory.[8] Ideally, to properly address the issue of the effects of technology, student assessments should provide data about performance both in traditional subject matters and in the more specific domain of the digital skills.

And finally, direct observations and other qualitative strategies are quite complicated methodological exercises that frequently bring up questions about intrusiveness or fake behaviors. However, such observations are the only means to understand in a situated way the dynamics of technology use in the classroom. They are essential for understanding the reasons behind teachers' professional choices, which would otherwise remain incomprehensible, as well as for knowing the rationale for students' preferences in detail. Ultimately, direct observations can result in a wide array of assessments of what works (Nachmias, 2004). When well documented, they facilitate the identification of the key factors to consider when replicating successful experiments and scaling up the lessons learned, thus setting the pace for true systemic innovation (OECD, 2010b). Additionally, qualitative strategies allow for cooperation, participation, knowledge-sharing (especially local knowledge), and respect. It suggests people are participants and not just beneficiaries of technology innovations (Lubin, 2016).

[8] This has been the case for developing models of technology acceptance that have been successfully tested empirically in the education sector (Davis, Bagozzi, & Washaw, 1989; Schwarz & Chin, 2007; Venkatesh, Davis, & Morris, 2007).

3 An Agenda for the Future

The research on the connections between technology, education, and development has significant weaknesses, but these are not solely due to the complexity of the field or failure to clearly formulate an appropriate research question. Unfortunately, research on educational technology seems to be subject to the same strains as any other area in education research. Ultimately, endeavors to conduct research focusing on collecting empirical evidence are still incomplete. As a result, some of the considerations expressed here, arising from research issues encountered when trying to establish connections between technology use and educational results, could, in fact, be predicted from the education research field as a whole. To avoid the same mistakes once again, it is critical to define a research agenda that responds to three clear challenges.

Firstly, in a field as complex as that of educational technology, controversy is more frequent than consensus (Bull, Knezek, Roblyer, Schrum, & Thompson, 2005). However, there seems to be widespread agreement on at least one point: there is a need for a body of organized and convincing evidence that contributes to both political and practical decision-making. This is the first challenge.

Secondly, the political climate has for decades been clamoring for more reliable accountability, to make returns on investment in education more transparent. In the specific case of technology investments in education, this creates a paradoxical situation: investments increase whereas evaluations continue to give inconclusive results.

Thirdly, there is the question of how research can produce sufficient evidence so that teachers see an opportunity to improve their instructional methods. It is not so much resistance to change as a calculation of the returns of the investment required (in terms of time, effort, and devotion): is the effort required for a teacher to wisely use technology in her instructional designs always going to provide good returns in terms of improving pupils' learning? This is the core issue, particularly in small developing countries, where the introduction of technology may pose additional problems to teachers rather than helping them to address their current needs.

Alternative models have tried to explain which factors influence the acceptance of technologies, which would involve changing well-established processes that are part of standard patterns of behavior or work. Of all these, the model which is most often used for empirical research in education is that of Davis et al. (1989). According to this model, there are two main factors that can predict whether a technological solution will be adopted or not and that have a decisive effect on the series of decisions that must be taken to put the solution into practice. These two factors are the perception of its ease of use (Kukulska-Hulme, 2007), strongly linked to the professional or personal skills required and, then, the perception of the relevance and validity of the solution (Landauer, 1995). In short, it is necessary to have a technical command of the proposed solution (for example, being able to set up a computer and make it work), but even more importantly, it is critical to have a clear understanding of the benefits that the solution will bring (for example, seeing how motivation for learning or, eventually, academic performance, can be improved). Without this positive understanding of its usefulness, or without the required skills to properly use it, the proposed technological solution will never be adopted.

According to the Davis model and others (Schwarz & Chin, 2007; Venkatesh et al., 2007), what makes people decide to modify their current professional practices by means of adopting a new technological solution is simply the expectation that they will achieve greater efficiency. Furthermore, the initial effort to be made, in terms of costs, training, or planning has to be compensated for sooner or later, otherwise, the cost-benefit equation would yield a negative result. The question is very simple, therefore: why accept a technological solution requiring additional effort if it does not eventually lead to greater efficiency or better results?

Thus, one of the major questions is the applicability of research results to teaching. But what are teachers concerned about? What are their needs in this area? In an original research project, a group of teachers was asked what would help them to benefit more from the results of research in their professional development (Van Velzen, 2013). The outcome was that the drivers of change, in the sense of promoting the adoption of new instructional strategies, were very obvious:

(a) There should be a demonstrable long-term effect on students' results.
(b) The proposed instructional method should contribute to making teachers' work more efficient.
(c) There is a wealth of students' experience with the methodology.
(d) There is a comparison with other methods, particularly those that are currently mainstream.

It would be a mistake to believe that the research community is unfamiliar with such concerns; quite the reverse. For example, in a survey conducted in 2005 among researchers on technology and education in the United States (Schrum & Glassett, 2006), opinions were collected on areas of work that the researchers saw as having received insufficient attention until that time. The outcome was the following list of outstanding issues, in order of importance:

(a) increasing studies on the impact of technology on teaching, learning, students' results, instruction, and instructional methods;
(b) investigating the role played by teachers' beliefs in the adoption of technology, effective teaching practices, school climate and culture, collaboration, and those with learning difficulties;
(c) considering issues of the technological base (the best software, the latest hardware, online learning, and nanotechnology);
(d) developing researchers' capacities;
(e) exploring the nature of the role of school managers and directors and of budgets; and
(f) determining the role of technology in special education in terms of content acquisition.

The problem is not a lack of agreement between researchers' priorities and the concerns of professionals and of policy makers, but rather the question of how to satisfy actual teachers' needs. This is the position argued by Roblyer (2004) when he proposes four types of impact research that could meet those requirements: analysis of comparative advantage, improvement of implementation strategies, monitoring of social impacts and, lastly, documentation of educational practices to measure their effects. The research agenda proposed by Alsop and Tompsett (2007) makes it easier to picture how practical research questions can be approached. Their proposal includes innovative elements; in a way, they follow the paradigm of research in the field of health. The idea behind this proposal is to move forward little by little from researching the effects of introducing a new component (an innovation) to ascertain whether it is worthwhile to continue, towards researching the collateral effects when this new component is spread to a large group. They basically propose that research, rather than trying to include everything in a single study, should pass through a number of stages so that it can ultimately complete a cycle which, at the end, would meet teachers' concerns and needs.

What Kozma was already arguing more than 20 years ago (Kozma, 1991) still holds true today: research should be centered on instructional designs rather than the technologies used. However, this is easier said than done. As well as the fact that the meaning of the word *technology* has evolved, the issues of the predominant paradigm in research on technology and education, outlined repeatedly by Clark (1994), have yet to be satisfactorily resolved. Tompsett (2013) recently attempted to apply this current view of analysis to the use of technology in learning programming languages and the result was terrible: few studies can be considered to be truly directed toward the production of evidence and, sadder still, what little evidence there is does not seem to be conclusive. Alsop and Tompsett (2007), in an interesting analogy between research and innovation in health and education, argued that research in the education field was overly focused on technological change, disregarding opportunities to show that certain technology-assisted innovations have educational benefits. If teachers are not sufficiently convinced of the benefits of adopting a technological change owing to lack of evidence, they will simply continue to use their current methods until it is proved that alternative pathways give better results with less risk and effort.

4 Concluding Remarks

Given the current state of the art, it is easy to conclude that decisions about technology use in education are far from being based on evidence. The limited scope and scarcity of the existing knowledge base would certainly support this conclusion. Moreover, in the absence of a robust knowledge base and appropriate monitoring and evaluation arrangements, there is no way to inform decision-making with empirical evidence. By prioritizing access to technology, policy makers convey a very simple message – that they are using taxpayers' money to modernize schools in a way that can be actually seen and touched. Yet, such a transformation can only happen if teachers are empowered, both technically and pedagogically, to benefit from this window of opportunity.

There is an even greater need to sustain technology-supported educational innovations in ways that contribute to making schools better suited to the needs and works of the network society. To do this in an effective way is not just a matter of financing, but even more of monitoring and assessing what works in education and disseminating that knowledge in ways that are meaningful for teachers and suitable for scaling up. The pending issue of how technology use relates to educational performance can be explored through correlations and will be done even more in the future, but accompanying empirical research and experiments will have to be carried out so as to build a useful knowledge-base. Rather than claiming that more research must be done, the conclusion is that research must be oriented differently.

It is well known that the adoption of an innovation essentially depends on end-users' perception of the advantage produced by the application of a new strategy in relation to what they are currently using (Rogers, 2003). Applying this principle to the concrete example of technology in education, it might be expected that technology can support the design of new strategies to do the following:

- learn better, for example, in a more personalized manner;
- learn more, that is, achieve better outcomes from learning; and
- learn different things, that is, achieve learning goals that only technology can make possible.

Research should clarify what strategies will make this possible and, in what conditions they would be feasible and ultimately capable of being applied widely. The feasibility issue is extremely important in the case of school learning because there may be many strategies that could turn out to be incompatible with the current configuration of schools and of the teaching profession.

Education research, relating to technology and also in general, is complicated by the nature of its subject and by the fact that contextual conditions limit its capacity to provide results from which generalizations can be made, thereby affecting its ability to contribute to the creation of theories that are universally valid. It is precisely for this reason that authors such as Berliner are convinced that social sciences and, in particular, education research are the true hard sciences to be cultivated (Berliner, 2002). These are not excuses but rather an attempt to explain why even the best research has so little impact in teaching and learning methods. In education, the problem of these so-called *ubiquitous interactions*, that is to say, the sheer number of variables that increase the difficulty of isolating impacts or combining the results of different studies, is omnipresent (Lederman, 2003). Any example relating to strategies to improve learning would serve – as it is very difficult to isolate the influence of instructional strategies – students' skills and abilities, their socioeconomic status, and their motivation as well as the interaction between all these variables.

And finally, there is a need for localized examinations of what is taking place in classrooms, particularly in the context of small developing countries. Despite that the international agenda for education seems to suggest that access may bring only benefits, there needs to be an understanding of local needs in local contexts. This is necessary in order to find more broad strategies that could be supported by technology, replicated, and appropriately scaled up. There is no such a thing as a universal technology-based solution to the current educational challenges of small developing states. Well-oriented, local research can help to recognize teachers as actors and not mere beneficiaries or users of well-packed technology solutions. No doubt, research has a role to play investigating further the role that technology solutions play in improving the quality of education. But the right research questions must be asked. Because educational phenomena are quite complex and multi-faceted, the right questions are not about whether or not to use technology, at all, but about which technology solutions can best suit the evolving learning requirements that each individual teacher has to manage in the classroom, considering the reality of teaching conditions and opportunities. Equipment may shine and speak by itself, but, unless it is properly embedded into sound teacher practices, no educational effects will be ever seen.

References

Alavi, M., & Leidner, E. D. (2001). Research commentary: Technology-mediated learning. A call for greater depth and breadth of research. *Information Systems Research, 12*(1), 1–10.

Alsop, G., & Tompsett, C. (2007). From effect to effectiveness: The missing research questions. *Journal of Educational Technology & Society, 10*(1), 28–39.

Ananiadou, K., & Claro, M. (2010). *21st century skills and competencies for new millennium learners in OECD countries*. Paris: OECD Publishing.

Arias Ortiz, E., & Cristia, J. N. (2014). *El BID y la tecnología para mejorar el aprendizaje: ¿cómo promover programas efectivos?* Washington, DC: Banco Interamericano de Desarrollo.

Banerjee, A., Cole, S., Duflo, E., & Linden, L. (2007). Remedying education: Evidence from two randomized experiments in India. *Quarterly Journal of Economics, 122*(3), 1235–1264.

Bebell, D., & O'Dwyer, L. M. (2010). Educational outcomes and research from 1:1 computing settings. *The Journal of Technology, Learning, and Assessment, 9*(1), 1–15.

Berliner, D. (2002). Educational research: The hardest science of all. *Educational Researcher, 31*(8), 18–20.

Berlinski, S., Busso, M., & Cristia, J. (2013). *The effects of technology use on learning outcomes in developing countries: A meta-analysis*. Washington, DC: Banco Interamericano de Desarrollo. Documento mimeografiado.

Bernard, R. M., Abrami, P. C., Borokhovski, E., Wade, C. A., Tamim, R. M., Surkes, M. A., & Bethel, E. C. (2009). A meta- analysis of three types of interaction treatments in distance education. *Review of Educational Research, 79*(3), 1243–1289.

Bernard, R. M., Abrami, P. C., Lou, Y., & Borokhovski, E. (2004). A methodological morass? How we can improve the quality of quantitative research in distance education. *Distance Education, 25*(2), 176–198.

Boster, F. J., Meyer, G. S., Roberto, A. J., Inge, C., & Strom, R. (2006). Some effects of video streaming on educational achievement. *Communication Education, 55*(1), 46–62.

Brynjolfsson, E. (1993). The productivity paradox of information technology. *Communications of the ACM, 36*(12), 67–77.

Bull, G., Knezek, G., Roblyer, M. D., Schrum, L., & Thompson, A. (2005). A proactive approach to a research agenda for educational technology. *Journal of Research on Technology in Education, 37*(3), 217–220.

Center for Promise. (2015). *Wired to learn: K-12 students in the digital classroom*. Washington, DC: America's Promise Alliance.

Cheung, A., & Slavin, R. (2013). The effectiveness of educational technology applications on mathematics achievement in K-12 classrooms: A meta-analysis. *Educational Research Review, 9*, 88–113.

Cheung, A., & Slavin, R. E. (2011). *The effectiveness of educational technology applications for enhancing mathematics achievement in K-12 classrooms: A meta-analysis*. Baltimore, MD: Johns Hopkins University, Center for Research and Reform in Education.

Clark, R. E. (1983). Reconsidering research on learning from media. *Review of Educational Research, 53*(4), 445–449.

Clark, R. E. (1994). Media will never influence learning. *Educational Technology Research and Development, 42*(2), 21–29.

Conlon, T., & Simpson, M. (2003). Silicon Valley versus Silicon Glen: The impact of computers upon teaching and learning: A comparative study. *British Journal of Educational Technology, 34*(2), 137–150.

Conn, K. M. (2014). *Identifying effective education interventions in Sub-Saharan Africa: A meta-analysis of rigorous impact evaluations*. New York, NY: Columbia University Academic Commons.

Cradler, J. (2003). Technology's impact on teaching and learning. *Learning and Leading with Technology, 30*(7), 54–57.

Cristia, J., Ibarrarán, P., Cueto, S., Santiago, A., & Severín, E. (2012). *Tecnología y desarrollo infantil: Evidencia del Programa Una Computadora por Niño. Documento de trabajo N° 304*. Washington, DC: Banco Interamericano de Desarrollo.

Cuban, L. (2001). *Oversold and underused. Computers in the classroom*. Cambridge: Harvard University Press.

Davis, F. D., Bagozzi, R. P., & Washaw, P. R. (1989). User acceptance of computer technology: A comparison of two theoretical models. *Management Science, 35*(8), 982–1003.

Dillon, A., & Gabbard, R. (1998). Hypermedia as an educational technology: A review of the quantitative research literature on learner comprehension, control, and style. *Review of Educational Research, 68*(3), 322–349.

Duflo, E., Glennerster, R., & Kremer, M. (2008). Using randomization in development economics research: A toolkit. In T. P. Schultz & J. Strauss (Eds.), *Handbook of development economics* (Vol. 4, pp. 3895–3062). Amsterdam, The Netherlands: Elsevier.

Evans, D. K., & Popova, A. (2015). *What really works to improve learning in developing countries? An analysis of divergent findings in systematic reviews* [Policy Research Working Paper 7203]. Washington, DC: World Bank Group.

Faber, B., Sanchis-Guarner, R., & Weinhardt, F. (2015). *ICT and education: Evidence from student home addresses* [NBER Working Paper No. 21306]. Cambridge, MA: National Bureau of Economic Research, Inc.

Fabos, B., & Young, M. D. (1999). Telecommunication in the classroom: Rhetoric versus reality. *Review of Educational Research, 69*(3), 217–259.

Fairlie, R. W., & London, R. A. (2012). The effects of home computers on educational outcomes: Evidence from a field experiment with community college students. *Economic Journal, 122*(561), 727–753.

Fairlie, R. W., & Robinson, J. (2013). Experimental evidence on the effects of home computers on academic achievement among schoolchildren. *American Economic Journal: Applied Economics, 5*(3), 211–240.

Falck, O., Mang, C., & Woessmann, L. (2015). *Virtually no effect? Different uses of classroom computers and their effect on student achievement* [IZA Discussion Paper No. 8939]. Bonn, Germany: Institute for the Study of Labor.

Friedman, W., Gerard, F., & Ralaingita, W. (2010). *International independent evaluation of the effectiveness of Institut pour l'Education Populaire's "Read-Learn-Lead" (RLL) program in*

Mali: Mid-term report. Research Triangle Park, NC: RTI International Council of Scientific Unions. Committee on the Teaching of Science.

Ganimian, A. J., & Murnane, R. J. (2016). Improving education in developing countries: Lessons from rigorous impact evaluations. *Review of Educational Research, 86*(3), 719–755. https://doi.org/10.3102/0034654315627499

Glewwe, P. (2002). Schools and skills in developing countries: Education policies and socioeconomic outcomes. *Journal of Economic Literature, 40*(2), 436–482.

Glewwe, P., & Kremer, M. (2006). Schools, teachers, and education outcomes in developing countries. In E. Hanushek & F. Welch (Eds.), *Handbook of the economics of education* (Vol. 2, pp. 945–1017). Amsterdam, The Netherlands: North Holland.

Glewwe, P., Kremer, M., & Moulin, S. (2009). Many children left behind? Textbooks and test scores in Kenya. *American Economic Journal: Applied Economics, 1*(1), 112–135.

Glewwe, P. W., Hanushek, E. A., Humpage, S. D., & Ravina, R. (2013). School resources and educational outcomes in developing countries: A review of the literature from 1990 to 2010. In P. W. Glewwe (Ed.), *Education policy in developing countries* (pp. 13–64). Chicago, IL: University of Chicago Press.

Global Education Monitoring Report. (2016). *Every child should have a textbook*. Paris, France: UNESCO.

Guislain, P., Zhen-Wei Qiang, C., Lanvin, B., Minges, M., & Swanson, E. (2006). *Overview: Information and communications for development – Global trends and policies*. Washington DC: World Bank.

Hattie, J. (2009). *Visible learning: A synthesis of over 800 meta-analyses relating to achievement*. London, England: Routledge.

Hikmet, N., Taylor, E. Z., & Davis, C. J. (2008). The student productivity paradox: Technology mediated learning in schools. *Communications of the ACM, 51*(9), 128–131.

Hsu, Y. C., Ho, H. N. J., Tsai, C. C., Hwang, G. J., Chu, H. C., Wang, C. Y., & Chen, N. S. (2012). Research trends in technology-based learning from 2000 to 2009: A content analysis of publications in selected journals. *Educational Technology & Society, 15*(2), 354–370.

Hwang, G.-J., & Tsai, C.-C. (2011). Research trends in mobile and ubiquitous learning: A review of publications in selected journals from 2001 to 2010. *British Journal of Educational Technology, 42*(4), 65–70.

Keser, H., Uzunboylu, H. S., & Ozdamli, F. (2011). The trends in technology supported collaborative learning studies in 21st century. *World Journal on Educational Technology, 3*(2), 103–119.

Kozma, R. (1991). Learning with media. *Review of Educational Research, 61*(2), 179–211.

Kozma, R., & Wagner, D. (2005). Core indicators for monitoring and evaluation studies for ICT in education. In D. Wagner (Ed.), *Monitoring and evaluation of ICT in education projects – a handbook for developing countries* (pp. 35–57). Washington, DC: infoDev.

Kremer, M., Brannen, C., & Glennerster, R. (2013). The challenge of education and learning in the developing world. *Science, 340*(6130), 297–300.

Krishnaratne, S., White, H., & Carpenter, E. (2013). *Quality education for all children? What works in education in developing countries*. New Delhi: International Initiative for Impact Evaluation (3ie).

Kukulska-Hulme, A. (2007). Mobile usability in educational contexts: What have we learnt? *International Review of Research in Open and Distance Learning, 8*(2), 1–16.

Lai, F., Zhang, L., Hu, X., Qu, Q., Shi, Y., Boswell, M., & Rozelle, S. (2012). *Computer assisted learning as extracurricular tutor? Evidence from a randomized experiment in rural boarding schools in Shaanxi* (Vol. 235). Stanford, CA: Rural Education Action Project.

Landauer, T. K. (1995). *The trouble with computers: Usefulness, usability, and productivity*. Cambridge, MA: MIT Press.

Lawless, K. A., & Pellegrino, J. W. (2007). Professional development in integrating technology into teaching and learning: Knowns, unknowns, and ways to pursue better questions and answers. *Review of Educational Research, 77*(4), 575–614.

Lederman, N. (2003). What works: A commentary on the nature of scientific research [editor's note]. *Contemporary Issues in Technology and Teacher Education, 3*(1).

Linden, L. L. (2008). *Complement or substitute? The effect of technology on student achievement in India*. Washington, DC: Infodev.

Lipsey, M. W., & Wilson, D. B. (2001). *Practical meta-analysis*. London: Sage.

Lubin, I. A. (2016). *Intentional ICT: Curriculum, education and development* [IBE Working Papers on Curriculum Issues No. 17]. Geneva, Switzerland: IBE-UNESCO.

Lucas, A. M., McEwan, P. J., Ngware, M., & Oketch, M. (2014). Improving early-grade literacy in East Africa: Experimental evidence from Kenya and Uganda. *Journal of Policy Analysis and Management, 33*(4), 950–976.

McEwan, P. J. (2015). Improving learning in primary schools of developing countries: A meta-analysis of randomized experiments. *Review of Educational Research, 85*(3), 353–394.

Means, B., Toyama, Y., Murphy, R., Bakia, M., & Jones, K. (2009). *Evaluation of evidence-based practices in online learning: A meta-analysis and review of online learning studies*. Washington, DC: U.S. Department of Education, Office of Planning, Evaluation and Policy Development.

Nachmias, R. (2004). Factors involved in the implementation of pedagogical innovations using technology. *Education and Information Technologies, 9*(3), 291–308.

OECD. (2010a). *Are the new millennium learners making the grade? Technology use and educational performance in PISA*. Paris: OECD Publishing.

OECD. (2010b). *Inspired by technology, driven by pedagogy: A systemic approach to technology-based school innovations*. Paris: OECD Publishing.

OECD. (2015). *Students, computers and learning: Making the connection*. Paris: PISA, OECD Publishing.

Olofsson, A. D., Lindberg, J. O., Fransson, G. R., & Hauge, T. E. (2011). Uptake and use of digital technologies in primary and secondary schools – A thematic review of research. *Nordic Journal of Digital Literacy, 6*(4), 208–226.

Pedró, F. (2012). *Connected minds. Technology and today's learners*. Paris: OECD Publishing.

Peslak, A. R. (2005). The educational productivity paradox. Studying the effects of increased IT expenditures in educational institutions. *Communications of the ACM, 48*(10), 111–114.

Power, T., Gater, R., Grant, C., & Winters, N. (2014). *Educational technology topic guide* [HEART topic guides]. London: The Health & Education Advice & Resource Team.

Reeves, T. C. (2000). *Enhancing the worth of instructional technology Research through "design experiments" and other development research strategies*. Paper presented at the Annual Meeting of the American Educational Research Association, New Orleans, LA.

Roblyer, M. D. (2004). *If technology is the answer, what's the question? Research to help make the case for why we use technology in teaching*. Paper presented at the Technology and Teacher Education Annual Conference, 2004, Charlottesville, VA.

Rogers, E. (2003). *Diffusion of innovations* (5th ed.). New York, NY: The Free Press.

Ross, S. M., Morrison, G. R., & Lowther, D. L. (2010). Educational technology research past and present: Balancing rigor and relevance to impact school learning. *Contemporary Educational Technology, 1*(1), 17–35.

Russell, T. (1999). *The no significant difference phenomenon*. Raleigh, NC: North Carolina State University Press.

Sandholtz, J. H. (2001). Learning to teach with technology: A comparison of teacher development programs. *Journal of Technology and Teacher Education, 9*(3), 349–374.

Schacter, J., & Fagnano, C. (1999). Does computer technology improve student learning and achievement? How, when, and under what conditions? *Journal of Computing Research, 20*(4), 329–343.

Schrum, L., & Glassett, K. f. (2006). *Research on educational technology: Challenges to implementation and impact of scientifically based research*. Paper presented at the American Educational Research Association Annual Meeting, San Francisco, CA.

Schrum, L., Thompson, A., Maddux, C., Sprague, D., Bull, G., & Bell, L. (2007). Editorial: Research on the effectiveness of technology in schools: The roles of pedagogy and content. *Contemporary Issues in Technology and Teacher Education, 7*(1), 456–460.

Schwarz, A., & Chin, W. (2007). Looking forward: Toward an understanding of the nature and definition of IT acceptance. *Journal of the Association for Information Systems, 8*(4), 230–243.

Solow, R. M. (1987, July 12). *We'd better watch out* [New York Times book review]. Retrieved from http://www.standupeconomist.com/pdf/misc/solow-computer-productivity.pdf

Tamim, R. M., Bernard, R. M., Borokhovski, E., Abrami, P. C., & Schmid, R. F. (2011). What forty years of research says about the impact of technology on learning: A second-order meta-analysis and validation study. *Review of Educational Research, 81*(1), 4–28.

Tompsett, C. (2013). On the educational validity of research in educational technology. *Journal of Educational Technology & Society, 16*(3), 179–190.

Triplett, J. E. (1999). The Solow productivity paradox: What do computers do to productivity. *Canadian Journal of Economics, 32*(2), 309–334.

van Rooij, S. W. (2009). Adopting open-source software applications in US higher education: A cross-disciplinary review of the literature. *Review of Educational Research, 79*(2), 682–701.

Van Velzen, J. H. (2013). Educational researchers and practicality. *American Educational Research Journal, 50*(4), 789–811.

Venkatesh, V., Davis, F. D., & Morris, M. G. (2007). Dead or alive? The development, trajectory and future of technology adoption research. *Journal of the Association for Information Systems, 8*(4), 267–286.

Winthrop, R., & Smith, M. S. (2012). *A new face of education. Bringing technology into the classroom in the developing world* [Working Paper]. Washington, DC: Brookings Institution.

Wu, Y.-T., Hou, H.-T., Hwang, F.-K., Lee, M.-H., Lai, C.-H., Chiou, G.-L., … Tsai, C.-C. (2013). A review of intervention studies on technology-assisted instruction from 2005–2010. *Educational Technology & Society, 16*(3), 191–203.

Zucker, A. A., & Light, D. (2009). Laptop programs for students. *Science, 323*(5910), 82–85.

Author Information

Francesc Pedró
UNESCO
Institutional address: 7, place de Fontenoy, 75007 Paris, France

Complete mailing address: 7, place de Fontenoy, 75007 Paris, France
Telephone number: +33145681017
Email address: f.pedro@unesco.org
Website: www.unesco.org

Francesc Pedró joined UNESCO's education sector in Paris (France) in 2011, where he leads the work on education policy. His multinational team looks after teacher policies and development, technology in education and education policies, including national policy reviews and comparative international research. From 2005 to 2011 he worked at the OECD Centre for Educational Research and Innovation (CERI). In his capacity as senior policy analyst, he was responsible for a number of projects including the *OECD Teaching and Learning International Survey* (TALIS) and the *New Millennium Learners Project*. He was also in charge of the reviews of *educational R&D* in OECD countries and of a project on *Systemic Innovation in Education*. His most recent publications include: *Are the New Millennium Learners Making the Grade? Technology Use and Educational Performance in PISA* (2010), *Inspired by Technology, Driven by Pedagogy* (2010), *Working Out Change: Systemic Innovation in Education* (2010), *Tecnología y educación: lo que funciona y porqué* (2011), *Connected Minds. Technology and Today's Learners* (2012), *Le numérique: une chance pour l'école* (2013), *Tecnología para la mejora de la educación: experiencias de éxito y expectativas de futuro* (2015), and *Digital Services for Education in Africa/Le numérique au service de l'éducation en Afrique* (2015).

Chapter 3
Technology for Education in Low-Income Countries: Supporting the UN Sustainable Development Goals

Daniel A. Wagner

Abstract Information and communications technologies (ICTs) are increasingly available, even in developing countries, leading to a number of pressing questions. Will ICTs for education (ICT4E) improve learning and educational quality? Might they increase the 'digital divide' with negative consequences for equity? What can ICT4E offer to help assist in the achievement of the new UN Sustainable Development Goals? The present review addresses these questions and uses of ICT4E in developing countries. First, a short overview is provided of how ICT4E intersects with the learning outcomes and the quality of education, including contemporary perceptions of what does and does not work in ICT4E. Second, an ICT4E framework is suggested as a way to reconceptualize the parameters currently in use that will lead to more effective ICT4E design solutions. Third, ICT projects in four key education subsectors – early childhood, basic as well as secondary education, and teacher education – help to disaggregate where and how interventions have been made in recent years. Finally, a set of investment domains in ICT4E is described, along with a set of specific suggestions to advance the field. The paper concludes with the suggestion that measurable, sustainable, and scalable design solutions in ICT4E are the best way to assist in achieving the UN Sustainable Development Goals.

Keywords Learning • Technology • ICT4E • UN SDGs • Framework • Design solution

D.A. Wagner (✉)
International Literacy Institute, Graduate School of Education, University of Pennsylvania, Philadelphia, PA, USA
e-mail: wagner@literacy.upenn.edu; http://www.literacy.org/

I.A. Lubin (ed.), *ICT-Supported Innovations in Small Countries and Developing Regions*, Educational Communications and Technology: Issues and Innovations,
DOI 10.1007/978-3-319-67657-9_3

1 Introduction

New technologies are of growing importance around the world and in many facets of everyday lives and livelihoods. These information and communications technologies (ICTs) may have special benefits for learning, both in and out of schools. At the same time, major claims are often made about the success of particular devices or applications before substantial research has been undertaken. This chapter explores the current state of ICTs for education (ICT4E), with a view of the past as well as recommendations for the future.

Evidence from low- and middle-income countries (LMICs) reveals that significant progress has been made toward international goals for education (UNESCO, 2015). Enrollment is up on average, and in many countries, gender parity is approaching equity. In addition, the types of educational resources found within many schools have evolved. ICT applications to promote learning are increasingly apparent, even in poor countries, leading to a variety of legitimate questions. Will ICT4E improve learning and educational quality? Might it increase the 'digital divide' with consequences for equity? What is the relevance of technology tools for improving educational assessments and management? These and related questions are bound up in the new United Nations Sustainable Development Goals (SDGs), ratified in 2015, and with a target date of 2030. This set of UN goals comprises 17 SDGs that increase the range of targets for international development. While there is no specific target on ICTs as a goal, technology is nonetheless mentioned in many of the specific goals, including in Goal 4, to *Ensure inclusive and quality education for all and promote lifelong learning.*

The use of new ICTs to improve education has been a topic of discussion as technology has evolved throughout the twentieth and twenty-first centuries. Radio, television, movies, computers, the Internet, mobile phones, and tablets have each been expected to 'revolutionize' learning processes and the way education functions in society. The review presented in this chapter considers uses of ICTs in the education sector, with a particular focus on early childhood, basic and secondary education, and on teacher training. While the spotlight here is on 'new' technologies, some older technologies are included if they provide useful evidence with implications for better understanding the field. In this review, the term *education* is broadly defined, including both children in school and out of school (including in preschools or non-formal education programs) as well as teachers. Further, the emphasis here is on *learning* – that is, not just access to schooling, but rather the educational *quality* that individuals take away from the experience of ICT4E, in terms of skills (e.g., reading, writing, math, and language) and attitudes (e.g., motivation and ambition) that may also change as a consequence.

With advances in ICT design and manufacturing, available devices have proliferated, and they now cost a fraction of earlier versions, with a great deal more power. This is especially visible in the growth of mobile phones, and increasingly, in tablet devices that are outpacing other forms of communications technology at an exponential rate. The present review is not comprehensive. It concentrates on learning and educational quality, rather than on the engineering of better devices, or the citing of so-called 'killer apps.' The review considers especially low- and middle-income countries (LMICs), including certain small countries, as well as marginalized populations within those countries; thus, the review does not provide a comprehensive look at the global use of ICTs for education.

The changing environment and growth of investment in ICTs have created a new reality, as overall funding globally has more than doubled over the past decade (World Information Technology and Services Alliance [WISTA], 2010). Some of the consequences for education are evident. First, it is now easier to access and experiment with multimedia and other electronic content for subject-matter instruction, both in and outside the traditional classroom setting. Second, data collection and feedback take significantly less time and effort. Third, it is easier to empower communities through ICTs by making information more accessible and transparent. Given these and other appealing benefits of ICT innovations for education, coupled with the diminishing costs of procurement, it is no surprise that the interest in ICT4E has grown rapidly.

What has not been clear is a conceptualization of the ways that ICTs can and should be used with different end-users in LMICs. This chapter provides such a structural perspective, along with heuristics for thinking about past, present, and future trends in using ICTs to support improved learning for those most in need.

2 ICTS for Learning and Education in Low-Income Countries

Held in 1990, the World Conference on Education for All in Jomtien (Thailand) was a watershed moment in international education and development. This conference embraced two key challenges: first, to significantly increase access to education of children in poor countries; and second, to promote the *quality of learning* in education. In 2000, at the Education for All conference in Dakar, these same two challenges were expanded in a more detailed set of education targets that were eventually integrated into the 2015 UN Millennium Development Goals (MDGs) and later into the 2030 UN SDGs.

Even before the UN goals, it was clear that the quality of education was a serious concern in low-income countries. For example, a World Bank national household survey in rural Bangladesh found that 3 years of schooling had approximately zero value in terms of learning achievement (Greaney, Khandker, & Alam, 1999). In other words, the effort of getting kids into school had little or no payoff with respect to educational gains. That was in 1999. Today, after nearly 18 years of substantial investments in education development, new studies are appearing with the same basic result: in many countries, children cannot read a single word, even after multiple years attending school (Gove & Cvelich, 2010; Seurat, 2012). Clearly, it takes more than just enrollment – rather, the need is to address the quality of learning.

A child's ability to read is commonly regarded as the principal indicator for assessing the quality of education, for both Organization for Economic Co-operation and Development (OECD) member states and LMICs. In fact, the three bilateral development agencies that gave the most aid in 2012, according to the OECD (2013), have written literacy into their official education strategies, either explicitly or implicitly. In most countries, inadequate reading ability in primary school is a strong determinant of future disadvantage in terms of educational, social, and economic outcomes. Furthermore, the global shift toward knowledge-based economies requires more literate workforces, and reading provides a crucial foundation for continuing education (Wagner, 2014a). It is for this reason that ICT4E initiatives have increasingly had reading/literacy as a focal point of attention.

The evidence is also growing that ICTs are changing the nature of children's learning and cognitive processes. Observational studies indicate that young learners (especially in wealthier communities with strong broadband access) often engage actively with websites, message boards, social media, and so on. If they have a choice, they may prefer to interact socially on the Internet on a mobile device rather than listen passively to an instructor or read a textbook. Researchers have described the reciprocal nature of learning using ICTs, e.g., the impact on and use of metacognitive skills (Terras & Ramsey, 2012) and the ways that ICTs have affected human discourse (Traxler, 2009). In other words, ICTs are changing both what is learned and how learning takes place. There is also increasing evidence that search engines (such as Google) are rapidly becoming tools that obviate the need to engage in memorization among both school children and adults – another way that ICTs are impacting human cognition (Carr, 2008; Sparrow, Liu, & Wegner, 2011).

Among the many ideas put forward for ICT4E is the notion that technology can help the poor (or poor countries) *leapfrog* into the twenty-first century (Servan-Schreiber, 1967). It is still a popular idea, regularly reiterated by ministers of education, technologists, and many educators. Thus, in order to consider, in broad strokes, the status of ICT4E today, it is important to begin by considering three contemporary claims concerning the technological revolution.

(i) Upside: Enthusiasm about ICTs and education In rhetoric that is familiar, the World Bank recently claimed, "ICTs can empower the lives of Africans and are driving entrepreneurship, innovation, and income growth" (World Bank, 2012, p. 5). A recent UNDP report stated, "These days, Web 2.0 and mobile technologies are being heralded for their revolutionary potential, not unlike the 1990s when the Internet's potential was only starting to be understood" (UNDP, 2013, p. 1). Such positive views are endemic in the international policy arena. How much credibility do they have?

Two iconic projects – *One Laptop Per Child* (OLPC) and *Hole in the Wall* (HIWEL) – are among the best-known ICT4E projects today among the lay public, decision makers, and even educational specialists. OLPC used brightly colored user-friendly laptops to appeal to young children, while Hole in the Wall created

Internet kiosks that comfortably accommodated the smaller stature of a child. In the period 2000–2010, each project captured worldwide attention, even though there was little or no scientific research or evaluations that confirmed the impact of either approach. OLPC (2013) sought to capitalize on the advantages of relatively low-cost and portable laptops (with some other features) to provide one-to-one computing for children. The design was mainly the laptop device itself; it seemed to be highly attractive and brought a great amount of media (and fiscal) attention. With respect to HIWEL (2013), Internet access was seen as the primary design approach, and little effort was made to create more than informal and user-friendly kiosks. Its purpose was to engage with children, and the project may or may not have succeeded in the goal of increasing Internet access, as credible data is scarce.

(ii) Downside: Disappointments about ICTs and education In spite of the majority view that ICTs are having (or will have) a major impact on education globally, and could help greatly in LMICs, there is a smaller but growing chorus of concern. From agency and donor specialists to academics and even to technologists in the private sector, there is concern that claims of success are premature. Also growing are the number of empirical reviews and meta-analyses that have called into question the 'hype' around adjectives like *radical*, *revolutionary* and *game-changing* technologies or apps that will change education 'as we know it' (e.g. Barrera-Osorio & Linden, 2009; Clarke, Wylie, & Zomer, 2013; Dodson, Sterling, & Bennett, 2012; McEwan, 2015; Wagner, 2005). Indeed, a more sober (and accurate) way of considering the role of ICTs in education is that technology can be an enabler or amplifier of other work already being undertaken.[1]

(iii) Critical side: Credibility gaps in ICT4E Concerns (and complaints) about what is credible evidence in education and, in particular, in ICT-based initiatives are widespread. Rigorous evaluation methods, such as randomized controlled trials (RCTs) have only recently been utilized in the ICT4E domain, and flaws mar many of these.[2] It is also important to remember that the rapid changes in ICTs over the past decade have made evaluation unusually difficult and prone to what some simply term failure. Often, by the time an evaluation study can be completed or even started, the ICT platform (phone, tablet, or software) may have changed so much that the study is no longer relevant. Such rapid change has made it harder to design and understand how to use ICTs for high-quality learning interventions. To date, ICTs for learning resembles other areas of educational reform: a fairly long period of trial and error, followed by slow and incremental changes as the research base develops. It is crucial to invest substantial time and resources in the initial design to create a *right-size* data management and evaluation system that addresses the *purpose* of the project (see the ICT4E framework below).

---

[1] See Clarke et al. (2013) on ICT4D as *enabler*; Toyama (2011) on ICT4D as *amplifier*, http://www.theatlantic.com/technology/archive/2011/03/technology-is-not-the-answer/73065/

[2] See Wagner, Castillo, Murphy, Crofton, and Zahra (2014) for an overview and critique of RCTs in education and development See Piper and Kwayumba (2014) in Kenya; Aker, Ksoll, and Lybbert (2012) for RCT on mobiles and literacy in Niger.

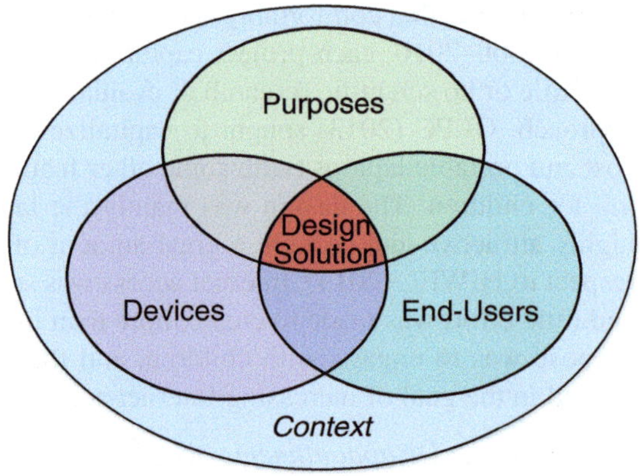

Fig. 3.1 ICT4E framework (Adapted from Wagner, 2014a)

In sum, the hope and excitement of new ICTs have been present from the inception of the United Nations Millennium Development Goals MDGs launched in 2000, initially with the creation of the UN ICT Taskforce. And, much has happened. Research, scholarship, and investments in ICT4E have grown substantially, overcoming initial resistance by some development agencies (and many ministers of education) that felt that the so-called 'information highway' was too expensive for poor countries that were 'still building ordinary roads.' Some agencies adopted ICT4E as a main focal point of their development efforts since 2000, such as in the work of IDRC-Canada (Elder, Emdon, Fuchs, & Petrazzini, 2014). New initiatives as part of the SDG discussion are appearing regularly. There is little doubt that major (even huge) investments will be made in ICT4E in the 2030 SDG period. The question remains: What should these investments be?

3 ICT4E Framework

With so many claims and counter-claims in the area of ICT4E, it is essential to find a way to conceptualize the kinds of initiatives that have been attempted or are in progress. One useful framework for considering interventions of ICTs for improving the quality of education includes four main components: purposes, devices, end users, and contexts (see Fig. 3.1).[3] Here, we identify six broad purposes through which ICTs can be used to support learning. These purposes intersect with devices and with types of end users to create a variety of design solutions, as described further below.

[3] There are other such frameworks: the FRAME model of Koole (Koole, 2006; Koole, McQuilkin, & Ally, 2010), which shares the notion of convergence but is less centered on design solutions. Also, there is a set of 'first principles' for ICT4E (Gaible, Bloome, Schwartz, Hoppes Poché, & Vota, 2011).

3.1 Understanding the Broad Purposes of ICT4E

Six primary purposes of interventions are described below, drawing principally on a recent landscape review of mobiles and reading, including 44 projects worldwide (Wagner, 2014a), with case examples provided.

(i) Formal learning and instruction This type of intervention is best characterized as intentional or directed learning. It incorporates guided inquiry through clearly planned learning objectives in a structured format, which the end user usually experiences through modules or units. Projects within this intervention purpose may concentrate on the acquisition of new knowledge and skills, such as phonemic awareness, vocabulary, or basic arithmetic. For instance, the Bridges to the Future Initiative (BFI) in South Africa, run by Molteno (2013), promotes reading acquisition through supplementary instruction using mobile tablets, as well as instruction supported by desktop computers (UNESCO Institute for Lifelong Learning (UIL), n.d.). An early evaluation of this computer-based approach was undertaken in India (Wagner, Daswani, & Karnati, 2010). Similarly, Ustad Mobile in Afghanistan (Paiwastoon, 2013) is a learning application that runs offline on simple feature phones and provides material in both Dari and Pashto from the first letter of the alphabet through grade-3 literacy and numeracy.

(ii) Informal learning Informal learning may cover a range of non-directed learning objectives, such as encouraging students to use SMS to practice reading or providing digital materials on mobiles in learners' local languages. Informal learning interventions offer learning materials or practices that try to engage the learners outside of formal learning settings. For example, the BBC Janala project (Walsh & Shaheen, 2013) in Bangladesh aims to teach English language listening and reading skills using a combination of interactive audio and mobile technology along with print and other ICT-based materials. In India, *VideoKheti* enables low-literate and novice users to find and watch agriculture-related videos through a text-free, multimodal interface with speech, graphics, and touch delivered through tactile smartphones and tablets (Cuendet, Medhi, Bali, & Cutrell, 2013).

(iii) Content Some interventions begin with the premise that the lack of textbooks leads to low-literacy environments that fail to support useful literacy practices, or that few text materials are available in locally written languages. For these reasons, interventions that provide e-books, pre-loaded with stories at varying levels of difficulty, are now becoming more commonplace. One such example is Worldreader, which distributed over 400,000 Kindle e-books to students and teachers in sub-Saharan Africa (UNESCO, 2014). Another project, Sesame Workshop's M is for Mobile initiative (Stewart, 2013), has begun to consider the potential of existing mobile technologies to reach educators and families that may have limited access to traditional media channels in India.

(iv) Training Training interventions contain direct or guided instruction for teachers who support children's learning – usually as part of teacher professional

development, but also may include facilitators or parents. The main purpose is to enhance the pedagogical skills of the teachers, most often by introducing them to new curricular or instructional content, such as lesson plans. For instance, in South Sudan, Across Radio uses solar-powered digital audio players to provide teaching resources and content in the local languages, Dinka and Bari. Recordings can also function as a supplemental literacy tutor for children in school or for children who are not in school. Education Development Center (EDC) has a long history of work with interactive radio instruction, e.g., in Zanzibar (EDC, 2009). More recently in Mali, EDC (2011) used Internet-enabled mobile phones to offer primary teachers access to resources and suggested lesson plans for French literacy instruction.

(v) Data collection Given the increasing emphasis on accountability in development work, some projects focus on rapidly assessing individual- and cluster-level learning gains in reading (Wagner, 2011). Such technologies aim to provide real-time monitoring of what a student does and does not know and to show teachers where to target their instructional efforts, so they can remediate the weaknesses they observe in learners. Some projects have back-end systems that can collect information on how the learning devices are used and can feed that information back into analysis and planning. Such systems usually collect a broad range of data from ICTs, such as the keystrokes used and participation time – both of which are difficult to deploy for real-time purposes. Two projects have allowed for real-time, student-level literacy assessment to identify individual strengths and weaknesses – one is *e-EGRA,* piloted in Mali and the Philippines (EDC, 2011–2012), and the other is *Tangerine*, piloted in Kenya and Indonesia (Research Triangle Institute (RTI) International, 2011).

(vi) Communication Some interventions attempt to harness the power of communications to engage in social exchange and quick delivery of information. Though not necessarily centered on learning, these efforts provide end users with the opportunity to engage in active dialogue, where learning and language skills are practiced informally. For example, the Pink Phone Revolution project in Cambodia uses mobile phones to empower women, by allowing them to participate fairly in the marketplace and to improve their safety and security (Mendoza & Vergel de Dios, 2012; Yardley, 2011). By using SMS messages, women receive current information on agriculture trends, market prices, and relevant disaster preparedness, along with cases of domestic violence. Similarly, the Yoza Project (2009–2013) in South Africa incorporates the use of mobile phones to communicate learning content and m-novels to create social media communities that foster literacy.

In sum, these six purposes provide one rubric to think about how project designers align their efforts, but these six are by no means exhaustive, nor are they guaranteed to persist in relevance as the ICT4E field continues to expand. In addition to purposes, we must think also in terms of three additional major parameters when creating a particular design solution. First, the *devices* themselves must be available, adaptable, and affordable. Second, *end user* characteristics must be thoroughly understood. *Third*, an adequate understanding of the *contexts* for implementation is crucial. Finally, at the intersection of the above parameters is what we term the *design solution*, as described further below.

(a) *Devices* Device-based implementations have been a hallmark of the history of ICT4E – from radio, movies, and television to mobile devices and the Internet cloud. Each of these advances has inspired both consumers and development specialists to imagine what would happen *if* a particular device were deployed. We take the position here that devices are necessary but not sufficient, in terms of appropriate and effective design solutions. For example, if local language content can be placed on a low-cost tablet, then it *might* help promote greater access to the practice of reading. If students are using SMS messages to communicate about their lives, then it seems reasonable that cell phones *might* offer an avenue that students, teachers, and volunteer facilitators could use to share information on school assignments or to provide advice on math or reading questions. Worldreader takes advantage of content provided in feature cell phones that have the greatest penetration in poor countries, making it available to those who do not have sufficient reading materials in their own local languages (UNESCO, 2014; Worldreader, 2013). This *might* help young readers, but we have little evidence as yet that it will. These are just three examples of how the device itself is a prominent component of the implementation but is insufficient without other solution components.[4] Devices play a critical but non-determinant role in ICT4E.

(b) *End users* In education, nothing is more important than the individual learner (whether in or out of school; whether a teacher or a student or manager of the educational system) – he or she is the ultimate end user. Educational interventions can and do fail when they are incompatible with particular characteristics of end users. And, conversely, ICT4E projects are more successful to the degree that they are designed with a particular end user in mind. For example, with respect to gender, ICT4E projects that included women's participation in the project design were generally more effective (Dodson, Sterling, & Bennett, 2013; Wagner, 2014a). Other interventions that have targeted specific ethnolinguistic groups by providing reading content in locally appropriate languages (or multi-lingual content) have found significant increases in student motivation for learning (Wagner et al., 2010).

(c) *Contexts* The context (and in particular, social, cultural, and linguistic context) is a key factor in any educational intervention. Thus, it is surprising that many projects provide few details on the *take-up* (in empirical terms) of an intervention within their particular contexts, such as in Hauser (2013) and Rangaswamy and Cutrell (2012). For example, if a project distributes tablets to students to use in the classroom for practicing local language literacy, how would this compare to the use of tablets at home outside of class hours, possibly with parents? Would these two contrasting contexts lead to differing results? Although project designers should keep contexts in mind as they choose the setting to implement an intervention, many developers seem (especially in ICT-based work) to ignore information on local context to inform the intervention design – perhaps thinking that, as with some ICT devices (the iPad comes to mind), users will adapt to the device, rather than vice versa.

[4] Additional useful resources on adapting devices to local contexts may be seen in Ho, Smyth, Kam, and Dearden (2009) and Medhi, Sagar, and Toyama (2007).

(d) *Design solutions* One general implication drawn from this landscape review is that it is possible to study the intersection of purposes along with devices and end users – what is characterized here as a *design solution*. Even so, it is not obvious from the programmatic and research reports available that the design solution is fully understood by the implementation teams *before* the project is put into place. In numerous instances, one or two parameters (e.g., the targeted end user and/or the technical parameters of the device employed) are left poorly specified.

When creating appropriate *design solutions* – whether started as device-based approaches or calibrated on a specific set of end users – projects face all the usual problems of any development intervention. More specifically, educational specialists confront the significant challenges of deploying different devices for different skills for different people at different ages in different contexts. This differentiation is particularly relevant to each subsector of educational interventions, as described in the following section.

4 ICT4E by Education Subsector

The significant growth of new investments in ICT4E gives the appearance of being haphazard – sometimes concentrated on a device (like OLPC laptops), or a particular population (such as refugees in South Sudan), or a private sector firm interested in expanding market share (e.g., Microsoft or Amazon). Indeed, some projects do converge on particular aspects of design solutions, as described above. Yet, the ways that the education sector provides resources (from local, national, and international agencies and organizations) can be understood, at least in part, by the subsectors that are age- and grade-related – even as it is clear that some design solutions span more than one subsector. Two clear examples would be: Bridges to the Future (BFI) in India (Wagner et al., 2010) that provides literacy support to young children as well as adults; and the Primary Math and Reading (*PRIMR)* Initiative in Kenya (Piper & Kwayumba, 2014) that supports early reading and teacher training.

(a) *Early childhood development* While policy interest in early childhood development (ECD) has grown in recent years, work on the role of ICTs in this area has been limited mainly to OECD countries (Bolstad, 2004; Kalas, 2010, 2012). Until fairly recently, television, such as in the well-known work of Sesame Street, has received the most attention and has been evaluated in terms of pre-school learning of the alphabet and other pre-reading skills.[5] The impact of Sesame Street has gone well beyond education, and now includes, in South Africa, outcomes related to HIV/AIDS prevention (Schierhout, 2005). Even so, the impact of such programs appears to be highly dependent on the level of exposure and on parental involvement

[5] For a recent overview, see Hinostroza, Isaacs, and Bougroum (2014). See ACER (2010) for work on Australian indigenous children. For international impact, see Cole (2009).

(Baydar, Kagitcibasi, Kuntay, & Goksen, 2008; Hinostroza et al., 2014). In such countries as Paraguay (Naslund-Hadley, Martínez, Loera, & Hernández-Agramonte, 2012), Bolivia, Honduras, Indonesia, Tanzania, and Malawi work using interactive radio instruction seems to have had a positive impact on the effectiveness of early childhood educators (Ho & Thukral, 2009).

With the advent of mobile devices (phones and tablets, in particular), a variety of new possibilities has opened up in support of the child's interaction with ICTs. These range from games that support early math and reading (Flannery et al., 2013; Gómez et al., 2013; Shuler, 2012), to data gathering for understanding children's learning (Pouezevara & Strigel, 2011). Some new efforts have been made to develop ways to send text messages to parents in support of home-based parenting skills, in ways similar to that used in the health sector (D'Sa, Borisova, & Nahis, 2014).

(b) *Primary education* Interventions in primary school have taken three main forms: (i) improved *access* to ICTs to schools (such as computers in labs, or the Internet); (ii) *device-assisted learning* (DAI) or direct instruction and consequential learning outcomes, mainly on small-scale basis; and (iii) improved *data capture* of classroom behavior of students and/or teachers.

With respect to ICT access, substantial resources have been spent on the use of ICTs in support of primary school children, with meta-analyses showing small but only occasional positive gains (Tamim, Bernard, Borokhovski, Abrami, & Schmid, 2011). Most of these studies reviewed in such reports were based on evidence from relatively wealthy OECD countries, where results were generally positive (OECD, 2012). However, studies in small developing countries that measured the impact of ICT access on student achievement showed mixed results: no positive impact on math and language skills in Colombia (Barrera-Osorio & Linden, 2009); compared to some positive impact on math test scores (with no effect on language test scores) in Ecuador (Carrillo, Onofa, & Ponce, 2010). A smattering of other project-specific studies showed some modest impact on learning in India and Chile, while a broader review in the United States showed mixed results.[6]

A number of studies have evaluated the *One Laptop Per Child* (OLPC) program, where laptops were distributed on a large scale, in a number of countries. Some, such as a study in Peru (Cristia, Ibarrarán, Cueto, Santiago, & Severín, 2012), found no positive effect on students' learning outcomes; another study in Uruguay (Ferrando, Machado, Perazzo, & Vernengo, 2012), had a positive impact on students' mathematics outcomes. Overall, while a great deal of attention and resources has been spent on the OLPC, little overall learning gains have been found. As with other interventions, these disappointing results may be the result of poor implementation as much as by the failure of this device-based approach. Indeed, relatively little attention seems to have been paid to the pedagogical aspects of OLPC as linked to the cultural contexts where it was deployed.

[6] On India, see Kam (2013) and Pawar, Pal, and Toyama (2006); on Chile, see Zurita and Nussbaum (2004); for a review of US studies, see Campuzano, Dynarski, Agodini, Rall, and Pendelton (2009).

A case study in Kenya A recent intervention by the Research Triangle Institute (RTI) in rural and peri-urban Kenya charted important new ground by investing in all four components of the framework design solution described above (in Fig. 3.1): (1) the primary purpose was to assess the impact of three types of ICT interventions on reading (via EGRA); (2) they employed three different device-type configurations (tablets for tutors, tablets for each teacher, and e-readers for students); (3) they emphasized specific end users (grade 2 children); and (4) they targeted a well-understood context of multi-lingual areas of Kenya, where RTI has worked for a number of years, and they trained a cohort of teachers in their methodology.[7]

Given the breadth and depth of this important and recent intervention, it merits detailed consideration. A wealth of data was gathered in an RCT design on over 1500 children across 80 schools, while measurement was limited to learning gains, and cost effectiveness. By comparing the three types of ICT-based interventions with control samples (who received no intervention), the researchers were able to demonstrate, across English and Kiswahili reading outcomes, a significant impact of ICTs for each of the ICT interventions when compared with baseline scores and with the control group. One further dimension of this work allowed for a comparison of cost-effectiveness among different ICT-based interventions; the most costly was that of the e-readers distributed to all children in that subsample, calling into question of oft-stated intention of some that one-to-one computing (i.e. each child has his/her own device) is optimal.

(c) *Secondary education* There are a number of important ways in which ICTs for secondary education differs from primary education, including a longer history of ICT investments; a better ICT infrastructure (e.g., stable electricity and computer labs); students who can work in international languages that are useful for ICT engagement; higher levels of teacher training around ICTs; and a closer rapport between ICT capabilities of youth and young adults and post-schooling work that increasingly requires individual ICT competencies. National case studies in Bangladesh (Khan, Hasan, & Clement, 2012) and Nigeria (Adomi & Kpangban, 2010) are among the many that support these points. Additionally, an up-to-date review of US secondary schools use of ICTs may be found in Leask and Pachler (2013).

There are many secondary education projects that have valued ICT infrastructure, teacher training, and learning skills for advancement to higher education. For the present review, we will consider three interrelated topics: open educational resources (OERs); the school-to-work transition; and non-formal youth education. OERs are "teaching, learning, and research materials in any medium, digital or otherwise, that reside in the public domain," as defined by UNESCO in 2002.[8] A number of international organizations and foundations have invested heavily in this area, particularly in support of secondary schools (Butcher, Moore, & Hoosen, 2014).

[7] Piper and Kwayumba (2014) for the PRIMR intervention study (cited with authors' permission). Note also that data collection was done via RTI's tablet-based EGRA assessment system, called Tangerine (Pouezevara & Strigel, 2011; Strigel & Pouezevara, 2012).

[8] http://en.wikipedia.org/wiki/Open_educational_resources

OERs are thought to support the teaching of twenty-first century skills by allowing students to explore the universe of knowledge made available on the Internet. This approach also connects to the notion that there is a school-to-work transition, whereby students can be better prepared for rapidly changing workforce needs. For example, Raftree (2013) has written on the use of mobiles in workforce development in LMICs and Levy and Murnane (2004) have tackled this same issue in industrialized countries. In a review of non-formal education initiatives for adolescent girls in more than a dozen developing countries, UNICEF (2013, p. 5) found that ICT could provide a "means for challenging unequal power relations and increasing participation of marginalized girls in social transformation." The methodology, deployed across countries, involved providing computer training and Internet access in community development centers with time reserved for this target population. Nonetheless, little evidence is provided on the effectiveness of these UNICEF-supported programs.

Overall, ICT use in secondary education is destined to expand, if only to meet the demands of workforce growth and development. Whether ICTs can help to radically change the way that pedagogy is organized for youth in and out of school remains to be seen.

(d) *Teacher education* Nearly a decade ago, Gaible and Burns (2005) produced a major review of the role of ICTs for teacher professional development (TPD) in developing countries, having consulted widely with specialists across the globe.[9] Their review included such areas as computers, the Internet, radio, television, and online distance learning. Even though new ICTs exist today, particularly in the form of mobile devices, the review covered many categories of current relevance, such as needs assessment, available resources, and monitoring and evaluation. In their analysis, each of the ICT approaches was considered in terms of its role in education, various strengths and weaknesses, and cost profile.

One area their review discussed in depth deserves further attention: the 'cascade model' of TPD. This model takes as its primary method the transfer of instructional capabilities from teachers/specialists who have more expertise, to those who have less expertise. A number of the projects highlighted in the Gaible and Burns (2005) review focused on how to strengthen elements of the cascade through TV, radio, and online TPD. One way to visualize the TPD cascade is shown in Fig. 3.2 (left side). The problem with the typical non-ICT cascade is that each teacher expert can only partially transfer skills to the next level down in the pyramid. ICTs are one way to 'flatten' the pyramid (shown on right side). One recent approach to this flattening effort involves the use of MOOCs (massive online open courseware) (Castillo, Lee, Zahra, & Wagner, 2015; Liyanagunawardena, Williams, & Adams, 2013; Wagner, 2014b). MOOCs have the potential to reach, with greater fidelity of transfer, a broader population of teachers who have access directly to top experts.

[9] See Unwin (2005) for a policy review with a concentration on Africa. An earlier comprehensive review with a focus mainly on the United Kingdom was undertaken by Cox et al. (2003). A later review of ICT and teacher professional development in the United States was done by Lawless and Pellegrino (2007). For a UNESCO perspective, see UNESCO (2008).

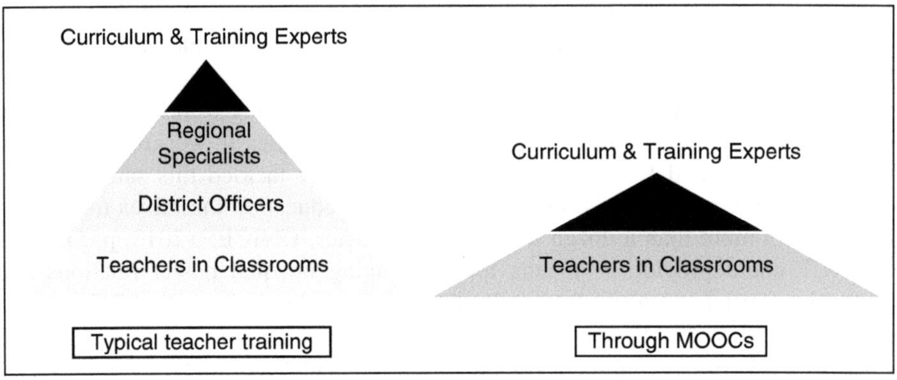

Fig. 3.2 MOOCs and Teacher Professional Development (Wagner, 2014b)

Besides MOOCs, other approaches are making themselves felt in the TPD arena, most notably through smartphones and tablets. Above, the RTI intervention in Kenya showed how tablets supported effective interventions in classrooms to support teachers; evidence is now coming in from other countries.[10] Overall, there is growing interest in using ICTs to support TPD. The general trend seems to be away from spending resources on training teachers *how* to use ICTs, and towards new ways to *support* what teachers already do in terms of instruction.

5 Investments for Supporting the UN Goals

There is no magic bullet (or single investment) in ICT for education any more than such a bullet exists in education more generally. Indeed, given the constantly changing landscape of ICTs, there is relatively little stable, well-researched knowledge that can, with confidence, pave the way forward. Nonetheless, as delineated above, there are a number of key domains that can help to guide thoughtful investments, those that have salience beyond the one-off claims of early success. While only notional at present, the recommendations provided below should help specialists and investors (private, public, non-profit) think about what would make sense in ICT4E in their own policy contexts.

(a) *Access and quality: The trend toward learning* One of the important transitions with the UN development goals is the shift from *access* to *quality* of education.[11] A parallel distinction may be seen in the ICT4E domain. In the early years, say from 2000 to 2010, there was the idea of pushing access to ICTs (whether through computer labs, kiosks with the Internet, and so forth). This notion continues today to a lesser extent with the idea that one can 'parachute' devices (e.g., Kindles, tablets,

[10] Tablet use in Australia (Pegrum, Howitt, & Striepe, 2013); mobile phones in Malaysia (Ismail, Bokare, Azizan, & Azman, 2013).

[11] http://en.unesco.org/post2015/

and smartphones) into villages. For the most part, research has shown that little learning impact has resulted from simple access projects. See the earlier discussion of *Hole in the Wall* in India, as a case in point. Clearly, the next decade or so (under the new UN SDGs) will inevitably be about investing in the quality of education, including the quality of ICT4E. This is a positive development that is supported by the available research literature and it supports the need to create optimized *design solutions* for learning, as described earlier.

(b) *Devices and content* As noted, devices are tempting, both among individuals (in homes) and among development agencies. We like them – they are glitzy, fast, exciting, and creative. But do they work for the goals we seek? That is the question. Generally, it must be conceded that projects with a unique focal point on devices are not likely to be successful or sustainable – if only because contextual factors are lacking and devices always make way for later devices. Nonetheless, devices that are well coupled with the broad factors in the ICT4E framework described earlier are more likely to succeed. As Traxler (2010) has rightly noted, there is a "tension between educational institutions aspiring to provide students with the technology for learning, ... and supporting students using their own devices" (p. 3).

Does the above assertion imply that learning content is more important than devices? Yes, generally speaking – assuming that the content is designed for specific end users and is relevantly adapted to cultural contexts. As ICT4E moves forward, we can observe a blending of devices and content, such as in the domain of OERs and MOOCs. When in doubt, we can conclude that it is better to invest in content over devices.

(c) *Costs and total cost of ownership* Until a decade ago, the perception was that ICTs are expensive and perhaps too expensive for LMIC development initiatives. Yet, surprisingly little information is provided publicly on the actual costs of devices, software, and human resources for implementation of projects. It appears that most ICT4E projects are funded by donor, government, or private sector agencies (or in some combination). This is predominantly a supply-side approach and perhaps appropriate in the initial stage of ICT4E work. Even the few projects that are integrated into a national curriculum seem to exist primarily because a donor agency has provided support to the education ministry to undertake the project.

To consider the cost implications of ICT4E, it is important to take into consideration the total cost of ownership (TCO). For example, in the Yoza Project in South Africa (Yoza Project, 2009–2013), students use mobile phones with a free app, but the cost of the social networking and texting is not free. It is unclear as yet whether the additional costs of using this social networking have a learning impact that merits the cost of the frequent usage. This cost comparison is similar to that of the TCO, where training and maintenance need to be added to the price of, say, a hardware device. It has also been suggested that costs can be reduced through a Bring Your Own Device (BYOD) approach, where learners can use their own mobile devices, with content loaded from a project implementer.[12] Equipment breakage is another key cost component. In Worldreader (2013), one notable impediment was

[12] For a discussion of TCO in developing countries, see Moretti (2011); for a similar discussion of the *total burden of assessments* in LMICs, see Wagner (2011). Few projects have tried BYOD approaches, but see one recent effort in Hong Kong (Wong, 2014).

that about 40 percent of devices broke during use and needed to be replaced during the one school year trial. Further, there are comparative costs, where variations of ICT inputs can have significantly different cost/learning impact (Piper & Kwayumba, 2014). Overall, cost is a significant barrier that can be met at present only by external assistance, but will no doubt be a key variable downstream. In any case, cost will be an element of growing importance for project sustainability.

(d) *Equity* As noted earlier, one of the main rationales for ICTs in low-income countries is the notion of being able to *leapfrog* to quality education. We have seen this rationale put forward by One Laptop Per Child (OLPC), Hole in the Wall (HIWEL), and some mobile devices distributed more recently. The rapid rise in mobile phone subscriptions in poor communities, while landline access has stagnated, tends to reinforce this perception. Yet, the current evidence does not support the idea of a leapfrog approach to quality education. In other words, increased distribution and/or use of ICTs does not translate (as yet) into better learning and school success among the poor.

It may be better to look at equity in terms of what some have called a *pro-poor* approach in ICT4E (Wagner, 2001, 2009). This approach puts the emphasis on the need to disaggregate the end user populations so that ICT4E becomes more effective among the very poor. Rather than implementing projects where there *already* is a device-friendly infrastructure, trained teachers, and electricity - equity necessitates moving into more challenging terrains. There is a need for measurable learning outcomes of the implementation among specifically targeted end users in poor communities. In other words, it is important to avoid simplistic 'trickle down' interventions that might later impact on poor children. Instead, find ways to disaggregate the target populations so that the solution design fits the purpose.

(e) *Public sector and private sector* The domain of ICT4E is unique in the public education sector in that it is necessarily joined with the private sector's ability to provide ICT devices (including Internet access) in support of learning. In other words, without the private sector, ICT4E would not happen. Further, when the public sector has, at times, sought to compete with the private sector (as when MIT sought to produce 'cheap' laptops in the OLPC initiative), the private sector simply produced cheaper and better products before OLPC could achieve scale. Similar examples abound, especially when claims for 'cheap' development tools were hyped as game-changers.

One of the challenges of device-based thinking among development specialists is whether they can or should get ahead of the privately-driven marketplace of new devices. The general consensus, based on the research findings to date, is that such non-market approaches largely fail. The market drives personal usage, such that the window of usability of purpose-built devices is rather short-lived (about 2–3 years). ICT4E programs that are built upon robust and well-populated devices (such as Nokia mobile phones, or Samsung tablets, or LCD projectors) are more likely to succeed long enough to establish a program of learning that merits the investment. A blend of public sector purposes with private sector device capabilities seems like the best path forward.

(f) *In-depth vs. scaling up approaches* One of the main attractions of ICTs in education is its potential scalability. As with 'apps' and Internet-driven commerce, marketing to many with a single tool is very seductive. With the right design solution, it is often argued, the ability to scale up reduces the per-person cost (and increases the return on investment) in direct proportion to the size of the denominator. This is one of the great attractions of MOOCs today, building on the success of earlier online and distance education systems. Naturally, such an approach only works to the degree to which the end user can, or is motivated to use a device (or website) that may be in a different language or dialect and/or provides information that is both relevant and potentially useful. If the education goal is to serve the bottom of the pyramid in LMICs, then issues of language, ethnicity, and relevance may be serious impediments to scalability.

The flip side is that of in-depth (ethnographically-centered) approaches to ICT4E. Even though some of these initiatives may not be scalable in the near-term, they may allow for a greater understanding of what might work in the future and provide models for best practice. Many of the projects reviewed in this chapter are not very scalable at present, as they are mostly in the pilot or project-based phase of work. Nonetheless, being able to go to scale – in some cases and in well-defined ways – is ultimately what will drive the use of ICTs in reaching development goals.

(g) *Design solutions* There is nothing more exciting than learning about the next 'big thing' in ICTs. Those who are old enough will remember the onset of emails, floppy diskettes, full-color Internet, 'high-speed' modems, mobile phones, USB sticks, and so on. With cutting edge devices and applications being announced so frequently, it is simply not possible to predict what new ICTs may be on offer in the coming few years. In Box 3.1, we recommend a number of key topics derived from the above ICT4E framework that could drive design solutions to impact the UN development goals.

6 Conclusions

The long-term implications for the use of new technology solutions are profound, both for the delivery of quality education and for a new vision of what it means to be an informed citizen in a world continuously transformed by technology. The prospect exists for ICTs to offer innovative tools that help meet the goals of improving education, reducing poverty, reaching gender equity, and improving child and maternal health. Nevertheless, we must admit that those same developments could create new disparities between rich and poor countries and communities.

The advent of ICTs in education offers an unprecedented opportunity, given their growing reach into populations across the globe: young and old, urban and rural, rich and poor, women and men. The challenges are many and inevitably complex. In this review, we have proposed an ICT4E framework that draws attention to the parameters of culturally appropriate and measurable design solutions.

In the domain of ICTs and learning today, there has been an extraordinary amount of publicity about the potential and claims of new technologies in developing countries. Some of these efforts are bearing fruit, but most remain in their infancy. We

Box 3.1 ICT4E suggestions for supporting the UN SDGs

- Where an installed base of computers already exists, provide local content to support reading instruction in multiple languages (including mother tongue) in primary schools.
- Implement and assess the comparative learning impact of differing ICT inputs in multiple sites in various countries.
- Carry out qualitative/ethnographic studies of the use of mobiles in the home and out of school to ascertain the ability of educational institutions to use these devices to support learning.
- Select one of the most disadvantaged populations in a LMIC (e.g., poor rural girls who are out of school) and create a design solution specifically for them – with adequate resources to recalibrate and overcome obstacles.
- Undertake a device inventory among the one or more marginalized populations in one or more LMICs, perhaps using graphical information systems technology.
- Undertake in-depth studies of the constraints of local capacity on success.
- Collect and use educational 'big data.' With the increase in digitally-collected learning outcomes, ministries of education can begin to detect local and regional systemic problems in instruction, closer to real-time decision making.
- Enable teachers to receive digital support information that is relevant to their instruction, and in real time. This might include instructional materials, advice from mentors, knowledge about individual student's learning progress, language support tools (in multilingual classrooms), and daily messaging to parents and siblings.

have argued that proclamations of success should be concentrated on outcomes and, in particular, on learning outcomes. The use of appropriate indicators is key to measuring and communicating the impact on student learning gains, teacher pedagogy, and accountability through community participation. Interventions with well-defined purposes can lead to real and meaningful advances in the knowledge base and in policy planning.

There is little doubt that the use of ICTs for improving education will continue to grow, along with ICT usage around the world. Investments in measurable, sustainable, and scalable design solutions, as part of a robust ICT4E framework, will be an important way to advance the field and learning and assist in achieving the UN Sustainable Development Goals.

Acknowledgements This paper is adapted from a review originally commissioned by the International Education Funders Group (IEFG), in association with the Firelight Foundation. The author would like to thank the following individuals for their time on the phone or for documents and/or comments received in connection with the IEFG review, including: Nathan Castillo, Amy Jo Dowd, Ed Gaible, Matthew Kam, Katie M. Murphy, Benjamin Piper, John Traxler, Mike Trucano, and Steve Vosloo. The views in this chapter are, however, the sole responsibility of the author, and not that of any person named above, or any organization or agency.

References

ACER. (2010). *Using television to improve learning opportunities for indigenous children*. Canberra, Australia: Australian Council for Educational Research.

Adomi, E. E., & Kpangban, E. (2010). Application of ICTs in Nigerian secondary schools. *Library Philosophy and Practice (e-journal)*, 345.

Aker, J. C., Ksoll, C., & Lybbert, T. (2012). Can mobile phones improve learning? Evidence from a field experiment in Niger. *American Economics Journal: Applied Economics.*, *4*(4), 94–120.

Barrera-Osorio, F., & Linden, L. L. (2009). *The use and misuse of computers in education: evidence from a randomized experiment in Colombia* [World Bank Policy Research Working Paper Series, No. 4836]. Available from http://elibrary.worldbank.org/doi/abs/10.1596/1813-9450-4836#

Baydar, N., Kagitcibasi, Ç., Kuntay, A., & Goksen, F. (2008). Effects of an educational television program on preschoolers: Variability in benefits. *Journal of Applied Developmental Psychology*, *29*(5), 349–360.

Bolstad, R. (2004). *The role and potential of ICT in early childhood education: A review of New Zealand and international literature*. Wellington, New Zealand: Ministry of Education.

Butcher, N. Moore, A., & Hoosen, S. (2014). Harnessing OER to drive systemic educational change in secondary schooling. *Journal of Learning for Development - JL4D, 1*(3), 2014.

Campuzano, L., Dynarski, M., Agodini, R., Rall, K., & Pendelton, A. (2009). *Effectiveness of reading and mathematics software products: Findings from two student cohorts*. Washington, DC: U.S. Department of Education, IES, NCEERA.

Carr, N. (2008, July/August). Is Google making us stupid? What the Internet is doing to our brains. *Atlantic Monthly*, pp. 1–8. Retrieved from http://www.theatlantic.com/magazine/archive/2008/07/is-google-making-us-stupid/306868/

Carrillo, P., Onofa, M., & Ponce, J. (2010). *Information technology and student achievement: Evidence from a randomized experiment in Ecuador*. Washington, DC: BID.

Castillo, N. M., Lee, J., Zahra, F. T., & Wagner, D. A. (2015). MOOCS for development: Trends, challenges, and opportunities. *Information Technologies and International Development*, *11*(2), 35–42.

Clarke, S., Wylie, G., & Zomer, H. (2013). ICT 4 the MDGs? A perspective on ICTs' role in addressing urban poverty in the context of the Millennium Development Goals. *Information Technologies and International Development*, *9*(4), 55–70.

Cole, C. F. (2009). *What difference does it make?: Insights from research on the impact of international co-productions of sesame street* [NHK Broadcasting Studies, 7]. Tokyo: Bunken (NHK Broadcasting Culture Research Institute).

Cox, M., Webb, M., Abbott, C., Blakeley, B., Beauchamp, T., & Rhodes, V. (2003). ICT and pedagogy: A review of the research literature. In *ICT in schools research and evaluation* (pp. 1–43). London, England: Department for Education and Skills.

Cristia, J. P., Ibarrarán, P., Cueto, S., Santiago, A., & Severín, E. (2012). *Technology and child development: Evidence from the One Laptop per Child Program* [IDB Working Paper IDB-WP-304]. Washington, DC: Inter-American Development Bank.

Cuendet, S., Medhi, I., Bali, K., & Cutrell, E. (2013). *VideoKheti: Making video content accessible to low-literate and novice users*. Paper presented at CHI 2013, Paris, France. Retrieved from http://research.microsoft.com/pubs/183489/Cuendet-CHI2013-VideoKheti-Final.pdf

D'Sa, N., Borisova, I. & Nahis, M. I. (2014). *Enhancing early child development by engaging fathers: Evaluation of a father support and education program in rural Bangladesh*. Paper presented at CIES Annual Meeting, Montreal, Canada.

Dodson, L., Sterling, S. R., & Bennett, J. R. (2012). Considering failure: Eight years of ITID research. *Information Technology and International Development, 9*(2), 19–34.

Dodson, L. L., Sterling, S., & Bennett, J. K. (2013, December). Minding the gaps: Cultural, technical and gender-based barriers to mobile use in oral-language Berber communities in Morocco. In *Proceedings of the Sixth International Conference on Information and Communication Technologies and Development: Full Papers-Volume 1* (pp. 79-88). Cape Town: ACM.

Education Development Center (EDC). (2009). *Radio instruction to strengthen education in Zanzibar: An impact study*. Washington, DC: USAID.
Education Development Center (EDC). (2011). *PHARE project*. Retrieved from http://idd.edc.org/projects/mali-usaidphare-program-program-harmonisé-dappui-au-renforcement-de-leducation
Education Development Center (EDC). (2011–2012). *eEGRA: Usable reading results-Right now!* Available from http://eegra.edc.org/
Elder, L., Emdon, H., Fuchs, R., & Petrazzini, B. (Eds.). (2014). *Connecting ICTs to development: The IDRC experience*. Ottawa, ON: Anthem/IDRC.
Ferrando, M., Machado, A., Perazzo, I., & Vernengo, A. (2012). *Impact assessment of the first OLPC national model: The Uruguayan Plan Ceibal*. Mimeo. Available from https://translate.google.com/translate?hl=en&sl=es&u=http://www.we-asc.org/alina&prev=search
Flannery, L. P., Silverman, B., Kazakoff, E. R., Bers, M. U., Bontá, P., & Resnick, M. (2013, June). Designing Scratch Jr: Support for early childhood learning through computer programming. In *Proceedings of the 12th International Conference on Interaction Design and Children* (pp. 1–10). New York, NY: ACM.
Gaible, E., Bloome, T., Schwartz, A., Hoppes Poché, J., & Vota, W. (2011). *First principles: Designing effective education programs using information and communication technology (ICT) – A compendium*. Washington, DC: USAID.
Gaible, E., & Burns, M. (2005). *Using technology to train teachers*. Washington, DC: infoDev.
Gómez, F., Nussbaum, M., Weitz, J. F., Lopez, X., Mena, J., & Torres, A. (2013). Co-located single display collaborative learning for early childhood education. *International Journal of Computer-Supported Collaborative Learning*, *8*(2), 225–244.
Gove, A., & Cvelich, P. (2010). *Early reading: Igniting education for all. A report by the Early Grade Learning Community of Practice*. Research Triangle Park, NC: Research Triangle Institute.
Greaney, V., Khandker, S. R., & Alam, M. (1999). *Bangladesh: Assessing basic learning skills*. Washington: World Bank.
Hauser, L. (2013). Qualitative research in distance education: An analysis of journal literature 2005-2012. *American Journal of Distance Education*, *27*, 155–164.
Hinostroza, J. E., Isaacs, S., & Bougroum, M. (2014). Information and communications technologies for improving students' learning opportunities and outcomes in developing countries. In D. A. Wagner (Ed.), *Learning and education in developing countries: Research and policy for the Post-2015 UN Development Goals*. New York, NY: Palgrave Macmillan.
Ho, J., & Thukral, H. (2009). *Tuned in to student success: Assessing the impact of interactive radio instruction for the hardest-to-reach*. Washington, DC: Education Development Center.
Ho, M. R., Smyth, T. N., Kam, M., & Dearden, A. (2009). Human-computer interaction for development: The past, present, and future. *Information Technologies and International Development*, *5*(4), 1–18.
Hole in the Wall Education Limited (HIWEL). (2013). Project website. Retrieved from http://www.hole-in-the-wall.com
Ismail, I., Bokare, S. F., Azizan, S. N., & Azman, N. (2013). Teaching via mobile phone: A case study on Malaysian teachers' technology acceptance and readiness. *The Journal of Educators Online*, *10*(1), 1–38.
Kalas, I. (2010). *Recognizing the potential of ICT in early childhood education: Analytica survey*. Moscow, Russia: UNESCO IITE.
Kalas, I. (2012). *ICTs in early childhood care and education: A policy brief*. Moscow, Russia: UNESCO IITE.
Kam, M. (2013). Mobile learning games for low-income children in India: Lessons from 2004-2009. In Z. L. Berge & L. Y. Muilenburg (Eds.), *Handbook of mobile learning*. New York, NY: Routledge.
Khan, M. H. S., Hasan, M., & Clement, C. K. (2012). Barriers to the introduction of ICT into education in developing countries: The example of Bangladesh. *International Journal of Instruction*, *5*(2), 61–80.

Koole, M. (2006). *Framework for the rational analysis of mobile education (FRAME): A model for evaluating mobile learning devices* (Unpublished Thesis). Centre for Distance Education, Athabasca University.

Koole, M., McQuilkin, J. L., & Ally, M. (2010). Mobile learning in distance education: Utility or futility? *Journal of Distance Education*, *24*, 59–82.

Lawless, K. A., & Pellegrino, J. W. (2007). Professional development in integrating technology into teaching and learning: Knowns, unknowns, and ways to pursue better questions and answers. *Review of Educational Research*, *77*(4), 575–614.

Leask, M., & Pachler, N. (Eds.). (2013). *Learning to teach using ICT in the secondary school: A companion to school experience*. New York, NY: Routledge.

Levy, F., & Murnane, R. J. (2004). Education and the changing job market. *Educational Leadership*, *62*(2), 80–83.

Liyanagunawardena, T., Williams, S., & Adams, A. (2013). The impact and reach of MOOCs: A developing countries' perspective. *eLearning Papers*, *33*.

McEwan, P. J. (2015). Improving learning in primary schools of developing countries: A meta-analysis of randomized experiments. *Review of Educational Research*, *85*(30), 353–394.

Medhi, I., Sagar, A., & Toyama, K. (2007). Text-free user interfaces for illiterate and semiliterate users. *Information Technologies and International Development*, *4*(1), 37–50.

Mendoza, D., & Vergel de Dios, B. (2012). *The pink phone revolution in Cambodia: A case study*. Phnom Penh, Cambodia: OXFAM.

Molteno (2013). Organizational website. Retrieved from http://www.molteno.co.za/our-programs

Moretti, M. (2011). Globalization of mobile and wireless communications: Bridging the digital divide. In R. Prasad, S. Dixit, R. van Nee, & T. Ojanpera (Eds.), *Globalization of mobile and wireless communications, Signals and Communication Technology* (pp. 19–29). Dordrecht, The Netherlands: Springer.

Naslund-Hadley, E., Martínez, E., Loera, A., & Hernández-Agramonte, J. M. (2012). *Leading the way to math and science success: Challenges and triumphs in Paraguay*. Washington, DC: IDB.

OECD. (2012). *Equity and quality in education: Supporting disadvantaged schools and students*. Paris, France: Author.

OECD. (2013). *Aid to poor countries slips further as governments tighten budgets* [News release]. Retrieved from http://www.oecd.org/newsroom/aidtopoorcountriesslipsfurtherasgovernments-tightenbudgets.htm

One Laptop per Child (OLPC). (2013). Project website. Retrieved from http://one.laptop.org/

Paiwastoon (2013). Project website. Retrieved from http://svr1.paiwastoon.net/mliteracy/

Pawar, U., Pal, J., & Toyama, K. (2006). Multiple mice for computers in education in developing countries. In *Proceedings of international conference on information technologies and development* (pp. 64–71). Berkely, CA: IEEE.

Pegrum, M., Howitt, C., & Striepe, M. (2013). Learning to take the tablet: How pre-service teachers use iPads to facilitate their learning. *Australasian Journal of Educational Technology*, *29*(4), 464–479.

Piper, B., & Kwayumba, D. (2014). *The Primary Math and Reading (PRIMR) Initiative, Kisumu Information and Communication Technology (ICT) Intervention: Endline report*. Washington, DC: RTI International.

Pouezevara, S., & Strigel, C. (2011). Using information and communication technologies to support EGRA. In A. Gove & A. Wetterberg (Eds.), *The early grade reading assessment: Applications and interventions to improve basic literacy* (pp. 183–226). Research Triangle Park, NC: RTI Press.

Raftree, L. (2013). *Landscape review: Mobiles for youth workforce development*. Washington, DC: USAID.

Rangaswamy, N., & Cutrell, E. (2012). Anthropology, development, and ICTs: Slums, youth, and the mobile internet in urban India. *Information Technology and International Development*, *9*(2), 51–63.

Research Triangle Institute (RTI) International. (2011). Tangerine project website. Available from http://www.tangerinecentral.org

Schierhout, G. (2005). *Impact assessment of a new programming component on HIV and AIDS*. Johannesburg, South Africa: Takalani Sesame Project, South African Broadcasting Corporation Ltd, Department of Education of South Africa and Sesame Workshop supported by USAID and Sanlam.

Servan-Schreiber, J. J. (1967). *Le Défi Américain*. Paris, France: Denoel.

Seurat, A. (2012). *Questions d'alphabétisation dans le contexte africain* (Unpublished doctoral thesis). Institute for Research in the Sociology and Economics of Education, Université de Bourgogne, Dijon, France.

Shuler, C. (2012). *iLearn II: An analysis of the education category of Apple's App Store*. New York, NY: Joan Ganz Cooney Center at Sesame Workshop.

Sparrow, B., Liu, J., & Wegner, D. M. (2011, July 14). Google effects on memory: Cognitive consequences of having information at our fingertips. *Science Express*, *333*(6043), 776–778.

Stewart, A. (2013). *"M" is for mobile: Sesame Workshop's mobile initiatives in India*. Presentation at Connected Living Asia Summit mEducation, Shanghai. Retrieved from http://www.gsma.com/connectedliving/wp-content/uploads/2013/03/MAE-Connected-Living-Summit_Sesame.pdf

Strigel, C. & Pouezevara, S. (2012). *Mobile Learning and Numeracy: Filling gaps and expanding opportunities for early grade learning* [Working Document. Prepared for GIZ]. Durham, NC: Research Triangle Institute.

Tamim, R. M., Bernard, R. M., Borokhovski, E., Abrami, P. C., & Schmid, R. F. (2011). What forty years of research says about the impact of technology on learning: A second-order meta-analysis and validation study. *Review of Educational Research*, *81*(1), 4–28.

Terras, M. M., & Ramsey, J. (2012). The five central psychological challenges facing effective mobile learning. *British Journal of Educational Technology*, *43*, 820–832.

Toyama, K. (2011). Technology as amplifier in international development. In *Proceedings of iConference 2011* (pp. 75–82). New York, NY: ACM.

Traxler, J. (2009). Learning in a mobile age. *International Journal of Mobile and Blended Learning*, *1*, 1–12.

Traxler, J. (2010). Will student devices deliver innovation, inclusion, and transformation? *Journal of the Research Center for Educational Technology*, *6*(1), 3–16.

UNDP. (2013). *ICTs and the post-2015 agenda: Learning from the sustainable networking programme*. New York, NY: Author.

UNESCO. (2008). *ICT competency standards for teachers. Policy framework*. Paris, France: Author.

UNESCO. (2014). *Reading in the mobile era: A study of mobile reading in developing countries*. Paris, France: Author.

UNESCO. (2015). *Education for All global monitoring report: Education for All 2000–2015: Achievements and challenges*. Paris, France: Author.

UNESCO Institute for Lifelong Learning (UIL). (n.d.). *Bridges to the future* [Project website]. Retrieved from http://www.unesco.org/uil/litbase/?menu=4&programme=166&language=en

UNICEF. (2013). *Integrating Information and Communication Technologies into communication for development strategies to support and empower marginalized adolescent girls*. New York, NY: Author.

Unwin, T. (2005). Towards a framework for the use of ICT in teacher training in Africa. *Open Learning: The Journal of Open, Distance and e-Learning*, *20*(2), 113–129.

Wagner, D. A. (2001). IT and education for the poorest of the poor: Constraints, possibilities, and principles. *TechKnowlogia: International Journal for the Advancement of Knowledge and Learning*. Retrieved from http://www.techknowlogia.org/TKL_Articles/PDF/304.pdf

Wagner, D. A. (Ed.). (2005). *Monitoring and evaluation of ICT in education projects: A handbook for developing countries*. Washington, DC: World Bank/InfoDev.

Wagner, D. A. (2009). Pro-poor approaches to using technology for human development: Monitoring and evaluation perspectives. In S. Bekman & A. Aksu-Koc (Eds.), *Perspectives on human development, family, and culture: Essays in honor of Cigdem Kagiticibasi* (pp. 367–380). London, England: Cambridge University Press.

Wagner, D. A. (2011). *Smaller, quicker, cheaper: Improving learning assessments in developing countries*. Washington, DC/Paris, France: EFA Fast Track Initiative and UNESCO-IIEP.

Wagner, D. A. (2014a). *Mobiles for reading: A landscape research review* [Technical Report]. Washington, DC: USAID/JBS.

Wagner, D. A. (2014b). *MOOCs for teacher professional development: Moving beyond the cascade method.* Presentation at Mobile Learning Week. Paris, France: UNESCO.

Wagner, D. A., Castillo, N. M., Murphy, K. M., Crofton, M., & Zahra, F. T. (2014). Mobiles for literacy in developing countries: An effectiveness framework. *Prospects*, *44*, 119–132.

Wagner, D. A., Daswani, C. J., & Karnati, R. (2010). Technology and mother-tongue literacy in southern India: Impact studies among young children and out-of-school youth. *Information Technology and International Development*, *6*(4), 23–43.

Walsh, C., & Shaheen, R. (2013). *English in action (EIA): Mobile phones as an agent of change for large-scale teacher professional development and English language learning in Bangladesh.* Paper presented at American Educational Association Annual Meeting, San Francisco, CA.

Wong, G. K. (2014). Engaging students using their own mobile devices for learning mathematics in classroom discourse: A case study in Hong Kong. *International Journal of Mobile Learning and Organisation*, *8*(2), 143–165.

World Bank (2012). *The transformational use of information and communication technologies in Africa* [e-Transform Africa report]. Washington, DC: Author & Tunis-Belvedère: African Development Bank.

World Information Technology and Services Alliance (WITSA). (2010). *ICT spending to bounce back by 2013* [Press release]. Retrieved from http://www.witsa.org/v2/media_center/pdf/2010Dplanet_PR_November-final.pdf

Worldreader. (2013). *iREAD mid-term results: All children reading in Ghana.* Retrieved from http://cdn.worldreader.org/wp-content/uploads/2013/10/Midterm-Results-Study.pdf

Yardley, H. (2011, March 8). *International Women's Day: Cambodia's pink tech revolution* [Web log comment]. Retrieved from http://www.oxfamblogs.org/southernafrica/?p=2246#sthash.Ip6bbFzf.dpuf

Yoza Project (2009–2013). Project website. Retrieved from http://yozaproject.com/about-the-project/

Zurita, G., & Nussbaum, M. (2004). Computer supported collaborative learning using wirelessly interconnected handheld computers. *Computers & Education*, *42*(2004), 289–314.

Author Information

Daniel A. Wagner
University of Pennsylvania
International Literacy Institute, Graduate School of Education

Complete mailing address: 3700 Walnut Street, Philadelphia, PA 19104
Telephone number: 215-898-9803
Email address: wagner@literacy.upenn.edu
Website (optional): www.Literacy.org

Dan Wagner is the UNESCO Chair in Learning and Literacy and Professor of Education at the University of Pennsylvania. He is Director of the International Literacy Institute, co-founded by UNESCO and the University of Pennsylvania, and Director of Penn's International Educational Development Program in graduate study. After an undergraduate degree in Engineering from Cornell University, and voluntary service in the Peace Corps (Morocco), he received his Ph.D. in Psychology from the University of Michigan, was a two-year postdoctoral fellow at Harvard University, a Visiting Fellow (twice) at the International Institute of Education Planning in Paris, a Visiting Professor at the University of Geneva (Switzerland), and a Fulbright Scholar at the University of Paris. Dr. Wagner has served as an advisor to UNESCO, UNICEF, World Bank,

USAID, DFID, and others on international development issues. He is a fellow of the American Psychological Association, the American Anthropological Association, and the American Educational Research Association. He recently served as Chair of the Brookings Global Research Task Force on Learning. Dr. Wagner was the recipient of the 2014 UNESCO Confucius International Literacy Prize. In addition to over 160 professional publications, Dr. Wagner has written/edited 25 books.

Part II
ICT-Supported Innovations in Policy and Practice

Chapter 4
ICT Curriculum Planning and Development: Policy and Implementation Lessons from Small Developing States

Sdenka Zobeida Salas-Pilco and Nancy W.Y. Law

Abstract It is becoming increasingly recognized that education solutions from one country, including ICT for learning and teaching in the curriculum, may not be suited to the context of another. It is probably best for each country to develop its own solutions or approaches based upon in-depth analysis of the local conditions and strategic priorities. However, there are many constraints that small developing states encounter, such as the availability of financial and human resources, and geographical isolation, which may limit their ability to develop, implement, and evaluate their own strategic development plans. This chapter examines the particular case of small states in Latin America and the Caribbean with a population under five million, such as Costa Rica, Uruguay, Panama, and Jamaica. We begin with a brief overview of the geographic and socioeconomic context of these countries, followed by a review of the history of ICT curriculum planning and development in these countries. The review shows that some of the challenges encountered are similar to those found in medium or large countries, and others that are more specific to small states. We then report on three case studies of ICT curriculum planning and development in the region that have very different histories: Costa Rica, Suriname, and the Caribbean Universities Project for Integrated Distance Education (CUPIDE). We conclude this chapter by identifying some of the lessons learned from our investigation that may be of value to policy makers and leaders concerned about effective strategies to support ICT curriculum planning and implementation in small states.

Keywords ICT in curriculum • Small states • Latin America • Caribbean • e-Learning strategic development

S.Z. Salas-Pilco (✉) • N.W.Y. Law
Faculty of Education, University of Hong Kong, Hong Kong, China
e-mail: sdenkasp@connect.hku.hk; nlaw@hku.hk; http://web.edu.hku.hk/staff/academic/nlaw

I.A. Lubin (ed.), *ICT-Supported Innovations in Small Countries and Developing Regions*, Educational Communications and Technology: Issues and Innovations,
DOI 10.1007/978-3-319-67657-9_4

1 Introduction

Information and Communication Technology (ICT) has become an important element on the education agenda of many countries since the last decade of the twentieth century. Digital literacy has been recognized as an important twenty-first century skill and the integration of ICT tools and resources to support learning have also been high on the agenda of many countries. Studies of ICT implementation have identified trends in ICT curriculum planning and development as well as challenges in their implementation. However, most of these studies are conducted in the developed world or countries with large populations. There are countries with populations around or under five million that have well-developed policies and practices in the area of ICT for learning and are well covered in the related research literature, such as Singapore, New Zealand, and three of the Nordic countries. On the other hand, it is difficult to find research studies or reports on ICT developments in school education in small states with relatively lower levels of economic development.

For more than a quarter of a century, many developed countries have emphasized the need for school education to leverage the potentials offered by ICT in order to advance the national education and development goals. The rationales put forward for introducing ICT in education (often now referred to as e-learning) are various, and important not only for large or developed countries, but possibly even more so for small countries, such as the potential that technology provides to break the constraints of geographic isolation and to improve access to high-quality open educational resources. This chapter takes as its focus the small states in Latin America and the Caribbean to examine what ICT curriculum planning and development has taken place in these countries, as well as the achievements and challenges encountered. We will also draw on this brief review to explore the success factors and lessons that can be drawn upon to support ICT curriculum developments in small states.

2 Small States in Latin America and the Caribbean

In this chapter, we define a small state as one with a total population of no more than five million and has membership in the United Nations. Based on this definition, there are 16 small countries in this region, six located on the American continent and ten in the Caribbean, with differing social, historical, and economic backgrounds. Most of these small states are members of the Commonwealth, with Suriname, Uruguay, Paraguay, and Costa Rica being the exceptions. Most of these countries have a population of fewer than 1.5 million inhabitants, with Saint Kitts and Nevis being the smallest with approximately 52,000 inhabitants. Guyana is the country with the lowest GNI per capita, US$ 4170, and The Bahamas has the highest GNI per capita, US$ 20,980 (CIA Factbook, 2015; The World Bank, 2015a).

Table 4.1 Size of the population and Gross National Income (GNI) per capita

Country	Population (millions)	GNI per capita (US$)	Adult literacy % pop.[a]	Internet users % pop.	Dominant language(s)
St Kitts & Nevis	0.05	14,490	98%	65%	English
Dominica	0.07	7070	94%	63%	English, French patois
Antigua & Barbuda	0.09	13,360	99%	64%	English, Antiguan creole
Grenada	0.1	7850	96%	37%	English, French patois
St Vincent & the Grenadines	0.1	6560	96%	56%	English, French patois
St Lucia	0.2	7080	90%	51%	English, French patois
Bahamas, The	0.3	20,980	96%	77%	English, Creole
Barbados	0.3	14,960	100%	77%	English, Bajan
Belize	0.3	4350	83%	39%	Spanish, English, Creole
Suriname	0.6	9470	96%	40%	Dutch, English, Sranang Tongo
Guyana	0.7	4170	88%	37%	English, Guyanese Creole
Trinidad & Tobago	1.2	15,550	99%	65%	English, Caribbean Hindustani
Jamaica	2.9	5220	88%	41%	English, English patois
Uruguay	3.0	16,350	98%	61%	Spanish, Portuñol
Panama	3.6	11,130	95%	45%	Spanish
Costa Rica	4.8	10,120	98%	49%	Spanish

Sources: World Bank (2015a), CIA Factbook (2015), International Telecommunication Union - ITU-UN (2015)

[a]Percentage of the population age 15 and above who can, with understanding, read and write a short, simple statement on their everyday life

Education development needs to be considered and understood within the broader social, cultural, and economic context of a given country. In the case of ICT in education, the general level of education and the ICT infrastructure and access of a country also matters (Hinostroza & Labbé, 2011). In this section, we will provide some relevant contextual background for this region before we examine further the central theme of this chapter. Table 4.1 shows the size of the population, the Gross National Income (GNI) per capita of each small country, the literacy rate, and the dominant languages spoken in each of the countries. Dominant languages spoken in a country is used as an indicator of the cultural diversity of the inhabitants of a country.

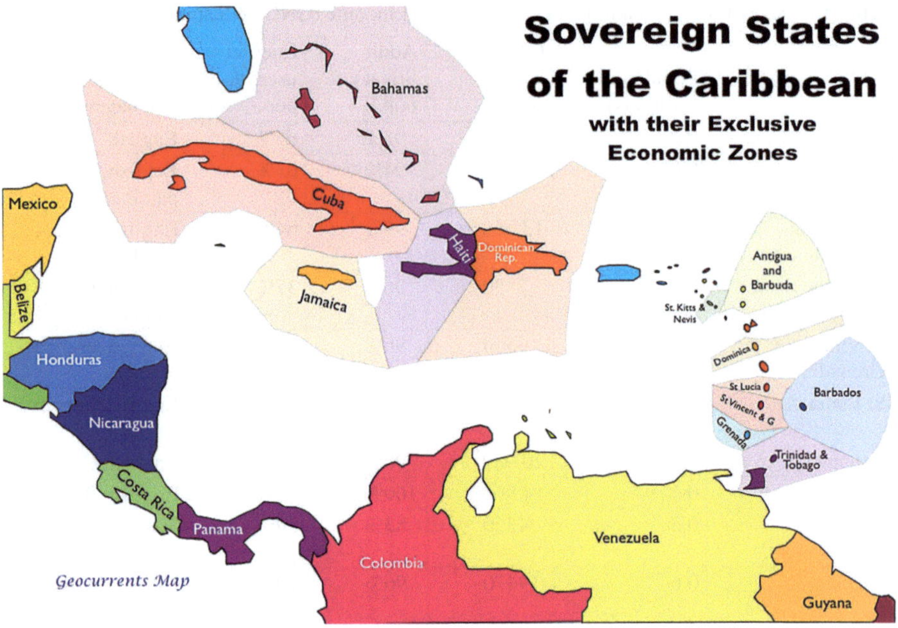

Fig. 4.1 Sovereign States of the Caribbean (Lewis, 2011)

2.1 *Geography*

The geography of this region is quite diverse and sometimes very challenging for transportation and communication, not only across countries but oftentimes also among the inhabitants of the same country. The challenges to development brought about by geographic isolation are more accentuated in the case of the small Caribbean islands (see Fig. 4.1 below). While new technologies may reduce the barriers due to geographic isolation, they do not serve as a panacea. Studies show that such potentials can only be unleashed by a highly educated population (e.g. Sachs et al., 2001; Wilke, 2010).

2.2 *History*

The history of a country or region is amongst the most important sources of contextual information, as it contributes importantly to the culture, language, religion, and other socioeconomic dimensions of a place. The American continent and the Caribbean were colonized firstly by Spain in the sixteenth century, which had a strong influence on the language and culture of the region. Later, other

European countries such as France, Britain, and The Netherlands colonized some territories in North, Central, and South America. The transatlantic slave trade, beginning in the sixteenth century, introduced people from the African continent to the region. Subsequently, during the nineteenth century, people from the Asian continent were brought in as indentured workers. This resulted in a rich diversity of ethnicities: Europeans mixed with local native indigenous populations of the Inca, Maya, and Aztec civilizations as well as other minority tribes in the continent, such as in the Amazon; and later persons of African and Asian ancestry. This cultural diversity has to be considered in order to avoid assuming a monoculture in making educational policy decisions (Henson, 2015; Parrish & Linder-VanBerschot, 2010). Further, e-Learning initiatives using online learning environments need culturally adaptive designs and implementation strategies in relation to language, values, and worldviews (Edmundson, 2011; Rogers, Graham & Mayes, 2007).

2.3 Population

While the populations of these 16 Latin American and Caribbean countries described above are small, the demographics of the populations in most of these countries are quite diverse. Some have populations that are ethnically almost homogeneous, such as the Bahamas, Barbados, Jamaica, Uruguay, and Costa Rica. In some others, such as Suriname, Trinidad and Tobago, and Guyana, the populations comprise a mix of European descendants from past colonization, indigenous communities, mestizos (mixed race), and descendants of Africans, who were brought as slaves by the European colonizers (CIA Factbook, 2015). The demographic background of the population shapes the contexts of these small states in terms of the relative homogeneity or diversity of languages, traditions, and worldviews within each country.

2.4 Economic Development

There are also big differences in economic development among the small countries in this region. Five of these countries are categorized as high-income, one as low-income, while the remaining nine as middle-income countries according to the World Bank definition (World Bank, 2015b). As can be seen from the data presented in Table 4.2, most of the small states included in the present work belong to middle-income economies with a population of less than half a million inhabitants. According to Crossley, Bray, and Packer (2009), globalization has had a big impact on the small islands, especially after the 2008 global economic crisis, because of their dependence on other countries' economies for their own development. This has also deepened the economic inequalities within these small countries and across them.

2.5 *Literacy Rate*

Most of the small Latin American and Caribbean countries have achieved universal primary education (Crossley, Bray & Packer, 2009), which is part of the Education for All (EFA) goals developed by UNESCO (2002). These goals aim to ensure access to quality education for the population in a country. In addition, these countries have a high percentage of adult literacy in their populations (see Table 4.1).

2.6 *Internet Access*

The Internet has the potential to reduce isolation and it gives access to rich open educational resources and courses, as well as the possibility to reach experts outside of the country at a very low cost. The figures in Table 4.1 indicate that in general, the higher income countries have higher percentages of their populations who are Internet users. However, income and Internet access rates are not strictly correlated. For example, Uruguay has the second highest GNI among the 16 countries but is ranked 7th in terms of the percentage of the population who are Internet users, at only 61%, according to the International Telecommunication Union (ITU-UN, 2015). On the other hand, Dominica is ranked 6th in Internet access at 63% of the population, while it is ranked only 12th according to its GNI. Clearly, there are more factors influencing Internet access than national income alone.

Table 4.2 Categorization by size of population and economic development

	Income categorization		
	Low-income	Middle-income	High-income
Population	GNI < $1045	$1046 < GNI < $12,735	GNI > $12,736
< 0.5 million		Belize Dominica Grenada St Lucia St Vincent & Grenadines	Antigua & Barbuda Bahamas, The Barbados Saint Kitts & Nevis
0.5–1.5 million	Guyana	Suriname	Trinidad & Tobago
1.5–5 million		Costa Rica Jamaica Panama	Uruguay

Source: GNI figures from The World Bank (2015b)

3 Policies and Implementation of ICT in the Curriculum

Most of the 16 small countries in Latin America and the Caribbean launched their ICT in education policies between 2000 and 2005 and implemented the planned programs between 2005 and 2010, showing the government commitments behind these policy initiatives (The Commonwealth, 2015; World Bank, 2011). One prominent exception is Suriname, a small Latin American country with a population of 600,000. It only launched its first policy related to ICT in primary education in 2014 (Ministry of Education and Community Development – Suriname, 2014a), and is the last country in Latin America and the Caribbean to have a policy on ICT in the school curriculum.

On the other end of the historical spectrum, Costa Rica stands out in terms of its early policy actions in this area. At the end of the 1980s, the Costa Rican government launched its first ICT related projects in schools, which introduced Logo programming in elementary schools to give children easy access to the concepts of computer programming as well as direct experience to do so. This project was supported by the Omar Dengo Foundation in close collaboration with the Massachusetts Institute of Technology (MIT). Costa Rica, a small Central American country, was thus among the first countries to implement an ICT curriculum project in primary schools and led the way ahead of many bigger countries not only in Latin America but also in the world (CEPAL, 1992).

Table 4.3 presents a list of the policy documents on ICT in education from the 16 countries that have been published until 2015. It shows that ten of these countries have already started to update and renew their policies in the second decade of the present century, based on the challenges experienced and the changing context since the launch of their first policy.

3.1 Nature of the ICT Curriculum Policies

Most of the policies listed in Table 4.3 focused on integrating ICT into the curriculum to support learning in all subjects, rather than on having a subject on IT skills. These were generally designed for implementation over a time span of three to five years. The early policies mostly targeted secondary education, and later included elementary education. Some countries even included special needs education in their ICT policies (UNESCO, 2012a).

These policies are largely concerned about four aspects of ICT in education: (a) ICT in curricula, (b) teacher training, (c) school infrastructure and connectivity, and (d) digital resources. Some of the policies, particularly those from the Caribbean islands, also included distance education (UNESCO, 2012a). It should be noted that distance education (DE) using radio and television is an older form of ICT, even though most people now associate ICT with the use of computers.

Table 4.3 List of policies, plans, and major initiatives on ICT in Education

Country	Date	Titles of published policies, plans, and initiatives
Antigua & Barbuda	2013	ICT in Education Policy
	2003	ICT in Education Policy
Bahamas, The	2009	10 Year Education Plan
	2003	Policy Statement on Electronic Commerce and the Bahamian Digital Agenda
Barbados	2010–2015	National ICT Strategic Plan
	2005	National Strategic Plan
	2000	Programme Edutech 2000
Belize	2011–2016	Education Sector Strategy
	2005–2010	Action Plan
Costa Rica	2010	Política para el aprovechamiento educativo de las tecnologías digitales
	2003–2015	Plan Educación para Todos
	2002	Programa Nacional de Informática Educativa
	1987	Convenio Marco de Cooperación MEP-FOD
Dominica	2004–2009	Implementing the National ICT in Education Policy
	2001	ICT Policy Framework for use in the Education System
Grenada	2006–2015	Revised Strategic Plan for Educational Enhancement and Development
	2001–2005	ICT Strategy and Action Plan
	2001	ICT4D National Policy
Guyana	2014–2018	Education Sector Plan
	2008–2013	Education Strategic Plan
	2006	ICT4D Guyana National Strategy
	2003–2007	Education Strategic Plan
	2001–2010	National Development Strategy
Jamaica	2011	ICT Policy
	2007–2012	National ICT strategy E-Powering
Panama	2009	Propuesta: Implementación de políticas TIC en la educación panameña
	2006–2010	Plan Estratégico Nacional para el Desarrollo de la Ciencia, la Tecnología y la Innovación

(continued)

Table 4.3 (continued)

Country	Date	Titles of published policies, plans, and initiatives
St Kitts & Nevis	2006	National ICT Strategic Plan
	2002	ICT Policy and Strategic Plan
St Lucia	2010–2015	National ICT Strategy
	2004	Integration of ICT in Education System
	2002	ICT Policy and Strategic Plan
St Vincent & the Grenadines	2010–2015	National ICT Strategy and Action Plan
	2002	ICT Policy and Strategic Plan
Suriname	2014	National Policy and Strategy on ICT in Education
Trinidad & Tobago	2014–2018	SmarTT – National ICT Plan
	2012–2016	SmarTT – National ICT Plan
	2003–2008	National ICT Strategy – FastForward
	2005	Draft Policy for ICT in Education
Uruguay	2008–2010	Agenda Digital (Plan Ceibal)
	2004	Project ITEEA (Integración Tecnológica al Entorno de Enseñanza y Aprendizaje)
	2000	Project INFED 2000

While the focus areas of these policies are similar, their visions are quite diverse. Many of these refer to goals related to the knowledge economy and use rhetoric that is commonly found in policy documents in developed countries. Such goals include developing knowledge (Panama), being ready for the 'knowledge revolution' (Antigua and Barbuda), improving the quality of life (Bahamas), preparing leaders for the knowledge society (Costa Rica). Some policies address specific local needs, such as being part of the poverty reduction strategy (Guyana).

It is important to highlight that ICT policy initiatives in these small countries started mostly as simple strategies, programs, or projects rather than clearly defined, system-wide policies. These were then followed by more formal ICT policies developed as part of a national plan. The list presented in Table 4.3 also shows this gradual policy evolution and development.

3.2 *Implementation: General Challenges*

Implementation of ICT-related policies necessarily involves ICT infrastructure provisions. This includes computational devices such as computers, tablets and smartphones, digital networks, and Internet connections. There are pronounced differences in the Latin American region, where some countries have higher indices of ICT infrastructure, such as Uruguay and Costa Rica, while others with relatively large rural and indigenous communities have very low indices, for instance, Ecuador and Suriname (Hinostroza & Labbé, 2011; The World Bank, 2015a). In addition, it is important to have technical support expertise for setting up and troubleshooting for the hardware and software used. The lack of necessary expertise could be an especially acute problem for island states with small populations (UNESCO, 2012b).

Most of the initial ICT projects in Latin America and the Caribbean were focused on equipping teachers with basic digital skills. This is usually the first step for introducing ICT in schools. However, these training initiatives were generally focused on technical skills and very few were focused on integrating them into pedagogical practices in the classroom (Lugo, 2010). This issue is not unique to small states or developing countries, but a general phenomenon reported in large-scale international studies as well (Law, Pelgrum & Plomp, 2008). The consequence is that even when teachers are trained in the use of ICT, they seldom use ICT in their pedagogical practices in schools (Buabeng-Andoh, 2012; Kozma & Vota, 2014).

Another problem often reported in the implementations of ICT curriculum policies is the general lack of proper evaluations (Trucano, 2012; World Bank, 2011). This is also the situation in most of the implementations in Latin American and Caribbean countries, irrespective of the size of the country (UNESCO, 2012b). Recently, international organizations, lenders, and donors have emphasized the importance of tracking the results and impact of ICT-related education project implementations, particularly as these usually involve high levels of investment. In addition to monitoring and accountability, appropriate evaluations will help to provide lessons learned from cycles of implementation to provide information to prevent future recurrence of the same mistakes. This could improve the efficiency and effectiveness of ICT integration in education, at the government, institutional, and individual levels (Trucano, 2012).

3.3 Implementation: Challenges that Are Acute for Small States

The challenges discussed in the previous section are not unique to small states but are generally found in the literature on ICT implementation in education. However, there are also challenges that are particularly acute to small states because of their population size. One such challenge is related to the lack of specialized personnel for planning and implementation (Crossley, Bray & Packer, 2009). Even in instances where there are funds to cover the high costs involved in recruiting foreign experts, they are often not familiar with local contexts.

Another issue for small countries is the difficulties in getting educational resources that are suitable for their local contexts. There are many Open Education Resources (OER) freely available on the Internet. However, these often need to be translated to the local language(s) even before any kind of adaptation can be considered. This translation and adaptation process again requires contracting specialized personnel and extra costs. One way these small states are overcoming this difficulty is to appoint experts who are multi-skilled to do multiple jobs (Butcher, Moore & Hoosen, 2014). For instance, by contracting experts with the capacity to carry out several different activities inside a project, it was possible for Guyana to implement their ICT initiatives at a low cost.

There are expectations that ready access to information and expertise through the Internet would make our world *flatter* (Friedman, 2005), and open learning opportunities brought about by recent MOOC developments is clearly one example of such developments. However, research on demographics of MOOC participants shows that most of the MOOC participants already have higher education qualifications or experience, located in industrialized countries, fluent in English, and are self-directed learners (Guo & Reinecke, 2014: Ho et al., 2015; Hood, Littlejohn & Milligan, 2015). The evidence so far is that current MOOC offerings have not contributed much to improving education access of small developing states, and the obstacle is not simply due to barriers related to ICT infrastructure or Internet access. To make significant contributions to educational developments in small developing states, MOOCs or other open learning provisions require tailoring of the courses to be relevant to the local needs and contexts, and appropriate policy support to build local capacities and expertise for the creation/adaptation of MOOC courses (Augustsson, 2014; Sucre & Garrett, 2016).

One important strategy to address the challenges of inadequate human resources and the need for adaptation is to develop regional collaborative support mechanisms. For example, Dominica, Grenada, and Saint Vincent and the Grenadines have followed the ICT Model Policy developed by the Organisation of Eastern Caribbean States Education Reform Unit (OERU), which provided cooperative support (Gaible, 2009).

3.4 Tackling Challenges Through Partnership and International Support

One way for small states to have a stronger voice is to join regional groups or international communities. Many of the small Caribbean island states belong to the Caribbean Community (CARICOM), the Commonwealth, and/or the United Nations. These memberships provide opportunities for them to receive support from more developed countries and international organizations in implementing ICT programs that require high levels of investment and expertise.

International aid and partnership for education development is another important source of expertise and funding that are strategically important for implementation success (Crossley, Bray & Packer, 2009). UNESCO has always played a crucial role in providing expertise and organizational support for teacher training programs in developing countries. The World Bank is also recognized for giving funds and providing ICT expertise. The Inter-American Development Bank (IDB), which is the main source of multilateral financing in Latin America, has contributed to the evaluation of some ICT-related education projects (Trucano, 2014). Another IDB initiative, launched in 2014, was the offering of free MOOC courses for government workers on development and sustainability. These courses are offered in Spanish, Portuguese and some in English, serving learners from Latin America and the Caribbean who are college educated and self-motivated (Inter-American Development Bank, 2016; Sucre & Garrett, 2016).

In the case of Uruguay, Costa Rica, and Panama, their governments have received support from UNESCO for planning in education. Their officials received the same training as those received by those from other large Latin American countries, through training programs organized by the International Institute for Education Planning (IIEP UNESCO), conducted in Spanish.

One challenge encountered by small countries in seeking aid is getting the funding agencies to 'think small'. Because of their population size, the scale of the projects and the funding sought may be relatively very small compared to other projects targeting large developing countries. Projects requesting such small amounts of funding could be sidelined because of the comparatively small perceived impact.

In addition to international organizations, some private companies such as Intel® and Microsoft® have also provided support in the form of training materials for teachers on basic digital skills as well as educational resources that teachers can use to support student learning in the classroom (Severin & Capota, 2011; UNESCO, 2012b). Partnership arrangements with international/regional organizations, private companies, and peer institutions are important to enhance access to resources and expertise in order to implement ICT projects in small developing states.

4 Country Case Studies

As mentioned in an earlier section, the context is of key importance for understanding the situation in small developing states. There are wide diversities across and sometimes also within the small states in Latin America and the Caribbean. In this section, we describe three case studies to illustrate how the ICT in curriculum planning and development may differ depending on the specific country context. The three selected cases were chosen because they differ greatly in terms of their context, experiences, and challenges encountered. In the case of Costa Rica, it was one of the first countries in the world to introduce ICT in schools and has an ethnically homogenous population. Suriname was the last country in the Caribbean to launch an ICT policy in education and has a very diverse population. The third case is a multi-institutional, multi-country collaborative ICT project in education implemented in several small states.

4.1 Costa Rica

Costa Rica is a small country located in Central America. The computers in education project launched in the late 1980s was among the first projects in the world to introduce ICT in schools. It started with a partnership between the Costa Rican government and the Omar Dengo Foundation (CEPAL, 1992; Muñoz et al., 2014). Hence, Costa Rica is a country with at least 30 years of experience in ICT implementation in education.

The initial general vision was for ICT use to stimulate children and the younger generation to develop new ways of behaving and thinking. Later this vision was expanded to include improving the quality of education, democratizing access to educational opportunities, and preparing citizens to move from an agricultural economy to more advanced modes of economic production. The Costa Rican policy development plan has some very important strengths: (a) the vision of leveraging ICT to foster higher level cognitive development through the introduction of new pedagogies, (b) collaboration with other institutions e.g. Omar Dengo Foundation, (c) maintenance of a long-term investment to sustain and increase the coverage of schools in the ICT program, (d) coordination and alignment of efforts in different areas to achieve better results in relation to objectives, (e) reduction in the digital gap through providing equitable access to ICT resources, and (f) improvement in the quality of secondary education. Their weaknesses are as follows: (a) the lack of a firm strategic policy, (b) lack of formal integration of ICT in the curriculum, (c) lack of connectivity in rural and elementary schools, and (d) lack of professional development for future teachers to use ICT in the schools (Fallas & Zúñiga, 2010).

In sum, Costa Rica has made advancements in expanding the percentage of primary and secondary students participating in the *Programa Nacional de Informática Educativa* (PRONIE-MEP-FOD) (see Fig. 4.2), as well as increasing the numbers of schools equipped with technology for teaching and learning (see Fig. 4.3). It can be seen that there were periods of sluggish development since the mid-80s, including the relatively long period from 2002 to 2012. It is heartening to see the great improvement since 2012, when the percentage increased almost 20% in 2 years, mainly through the provision of mobile technology (see Fig. 4.3). As of 2014, 80% of the student population had access to ICT in their schools.

4.2 Suriname

Suriname, a small island country located in South America, did not have any policy on ICT in education until 2014 (Suriname Ministry of Education and Community Development, 2014a). It is one of the last countries in the world to develop and implement an ICT policy.

The policy vision is for ICT use to bring equal opportunities to individuals to develop their talents effectively. The government's plan was to kick-start the

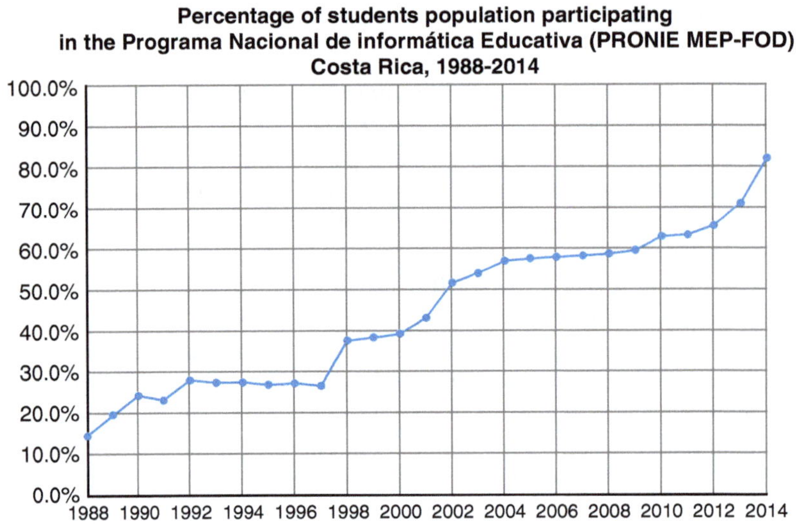

Fig. 4.2 Percentage of students in the program PRONIE MEP-FOD, 1988–2010 (Adapted from Muñoz et al., 2014, p. 93)

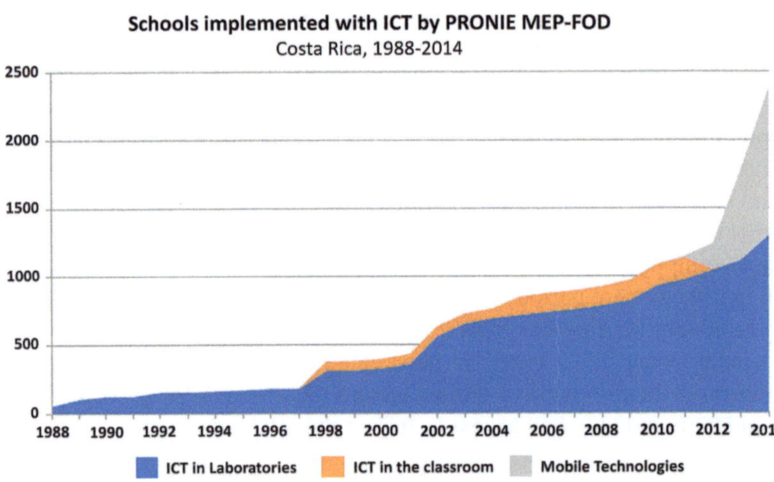

Fig. 4.3 Number of schools implemented with ICT by PRONIE MEP-FOD.1988–2014 (Adapted from Muñoz et al., 2014, p. 84)

implementation using the national budget and to seek financial support for further investment in the ICT initiatives. The case of Suriname alerts us to the fact that many countries are still struggling to develop and implement their first ICT in education policies.

The government of Suriname has shown a strong commitment to supporting the use of ICT in education to improve the learning process. However, there are still insufficient resources to carry out the proposed ICT strategy. Only a few pilot projects have been planned for the policy implementation (van der Hijden, 2014).

A survey carried out in 2013 shows the ICT situation in Suriname. Approximately 69% of pre-primary and primary schools have reliable access to electricity (see Fig. 4.4 below), 48% of schools have computers acquired on their own, but they are used mostly for administrative activities (see Fig. 4.5). Only 33% of schools have Internet access, teachers do not have competencies for using ICT, and schools located in the interior of the country have less access to ICT tools (Suriname Ministry of Education and Community Development, 2014a). It is important to highlight that computers and computer labs, where available, were acquired by the schools through their own initiatives. Inadequate technological infrastructure; low access to ICT in schools; and lack of human, financial and technological resources are still major challenges, especially in the interior of the country (Suriname Ministry of Education and Community Development, 2014b).

The difficulties and challenges faced by Suriname in introducing ICT in education are similar to those encountered by other small developing countries. It also demonstrates that the government's policy commitment is a necessary first step, that has to be followed by the development of mechanisms to raise funds for carrying out the implementation plan, which would be more challenging than those in developed countries, as it would have to include funding to (1) improve the basic technological infrastructure of the country, such as the provision of a stable supply of electricity to schools; and (2) support teacher training that encompasses basic ICT use. The biggest challenge would be to ensure that the implementation provides equitable and inclusive opportunities to schools located in remote areas and for marginalized communities. The case of ICT in education development in Suriname also demonstrates the strength of bottom-up grassroots support. Some schools in Suriname, without the government's financial support, took the initiative to equip themselves with computers and used them for teachers' in-service training and for developing pupils' ICT skills (Suriname Ministry of Education and Community Development, 2014a). This shows that the people will find solutions and act in autonomous ways to advance the educational provisions of their children even when governments do not take the necessary policy actions. On the other hand, it also demonstrates that governments could build strategic partnership with the local communities in their policy implementations, in addition to seeking aid and partnership with external/international institutions.

4.3 Multi-country Collaborative ICT Project in the Caribbean

This third case study provides another model of strategic development for ICT curriculum planning and implementation in small countries. The Caribbean Universities Project for Integrated Distance Education (CUPIDE) used satellites for Distance Education (DE), and was a successful collaborative project involving five Caribbean universities: The University of the West Indies (UWI), the University of Technology

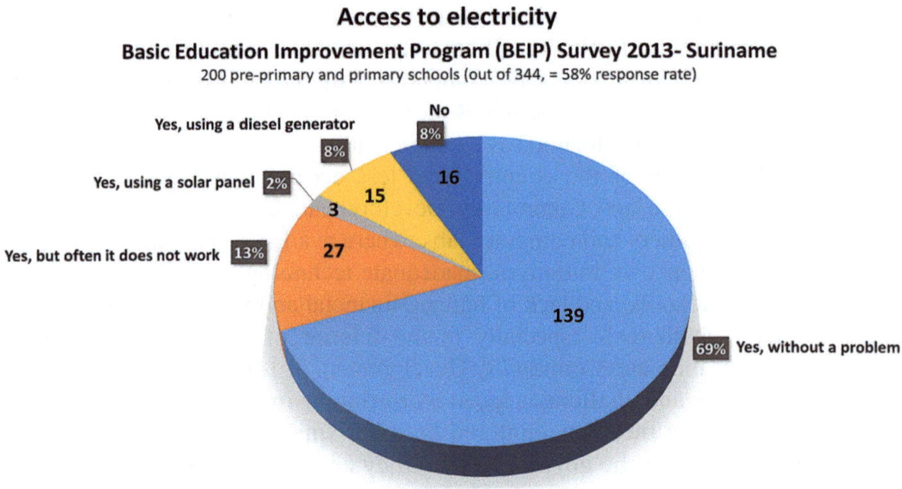

Fig. 4.4 Number of pre-primary and primary schools with electricity (Adapted from Suriname Ministry of Education and Community Development, 2014b, p. 4)

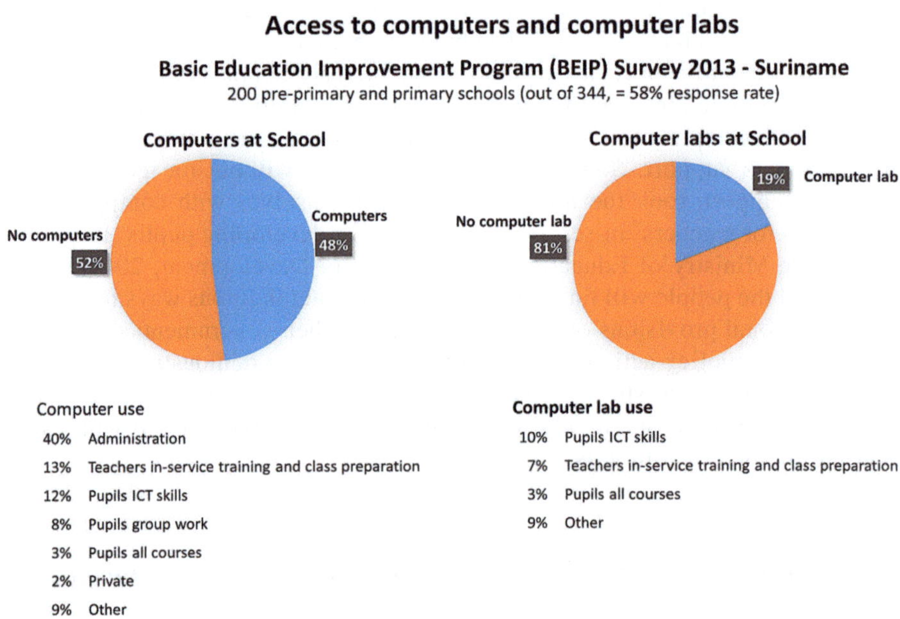

Fig. 4.5 Percentage of computers and computers labs in pre-primary and primary schools (Adapted from van der Hijden (2014))

of Jamaica (UTech), the Anton de Kom University of Suriname (AdeKUS), the University of Guyana (UG), and the University of Quisqueya (UniQ), Haití.

The project began in 2003 and was completed in 2007, by which time the project had offered in total 18 online and blended courses and had trained more than 130 academic staff, approximately 40 technical staff, and 20 student leaders. The project strengthened the links among the participating universities, helped to develop needed human resource expertise with the support of consultants, and fostered a multi-country collaboration. One major difficulty encountered was getting an affordable satellite bandwidth. Some other challenges that had to be overcome during the process of implementation included: (a) organizing regular meetings of individuals who were located at far distances from each other, to coordinate and make decisions for this multi-country project. This was accomplished using teleconference and face-to-face meetings; (b) accommodating for the language diversity of the participating countries by organizing planning and training events in three different languages: English, French, and Dutch; (c) customizing the training to meet the learners' needs from different countries; and (d) maintaining a detailed documentation of events and processes because of frequent staff changes (UNESCO, 2012b).

The CUPIDE project is considered successful in helping to develop the infrastructure and human resources of the five participating countries (UNESCO, 2012b). In addition, it has been revamped for the period from 2013 to 2017, and it now includes more small states such as Antigua and Barbuda, Belize, Dominica, and Grenada, among others (Government of Canada, 2015). This case is a good example of how small countries, or institutions across small countries, can form collaborative consortia to work together on specific initiatives that address identifiable common needs. By working together, they can pool and make use of their limited resources more effectively, as well as become more visible in attracting partnership arrangements with the private sector and in gaining support from international agencies to achieve long-term developmental goals.

4.4 *Lessons Learned from the Three Case Studies*

Taken together, the above three cases capture some of the important diversities in context as well as in the strategies taken by small countries in Latin America and the Caribbean in relation to ICT policies in education. Some insights on policies and strategies for ICT to support educational development in small countries can be gained from an examination of these three cases.

One clear observation across the three cases is the importance of vision and leadership at the policy level. Costa Rica was able to engage in pioneering implementation of ICT for learning 30 years ago as a small, developing country because of the foresight and vision of the country's leadership. The launch of the first National Policy and Strategy on ICT in Education in Suriname in 2014 was accompanied by state-level investment for the implementation, even though the focus had to be on activities that were achievable within the country's existing budgets rather than full-scale implementation. CUPIDE would not have been possible without the concerted commitment of the leadership from the five universities to jointly realize the vision of making use of satellite and digital technologies to serve the distance education needs of their respective countries.

A second observation is that some countries are taking advantage of mobile technology as a main form of ICT infrastructure provision for schools, such as in Costa Rica. Besides lowering costs, mobile technology has the added advantage of affording Internet access in the school without requiring the investment of wired Internet connected to and within buildings. Mobile technology also has the advantage of supporting student-centered learning and outdoor, in-situ learning activities. In addition, recent advances in solar-powered batteries make it possible for communities to use these batteries to charge the mobile devices even in communities that do not have access to a stable electricity supply.

Thirdly, for policy vision and goals to be realized, they must be accompanied by strong implementation plans and strategies that include appropriate resourcing, partnership support, and collaboration. CUPIDE had regular meetings of the leadership team members to address several implementation issues, which were many and complex, even when the goals were relatively specific and tangible. The Surinamese national ICT in education plans could not be effectively implemented without addressing the fundamental issues of electricity supply as well as computer and Internet access.

Another strength of the CUPIDE project was its focus on building the human resource capacity as a core implementation strategy. The project not only trained academic staff involved in delivering the courses, but also technical staff as well as student leaders. This allowed the project to build up its pool of expertise in the collaborating universities for sustainable operation and further expansion.

5 Conclusion

Nowadays, all governments understand that ICT policies are important for their country's future development and that ICT use should be integrated into the school curriculum. Technology is advancing and changing our societies, and policy makers and educators need to respond to the changing demands of education. This poses challenges to large and small countries alike. Yet, small countries, particularly those that are economically less developed and culturally more diverse, face even greater challenges in the planning and implementation of ICT use in the curriculum. Countries that do not already have a strong technological infrastructure, such as reliable electricity supply and Internet availability would have to make appropriate adaptations to ensure that their implementation plans are feasible. While fostering students' capacity for knowledge creation is important for countries to be able to engage in a knowledge economy, the educational goals for ICT implementations also need to be compatible with the overall level of literacy and economic development of a country. Further, the potential contribution of new educational and human development opportunities that are made available through open learning technology developments such as MOOCs can only be realized if these can be customized to serve the contexts and needs of the small developing states.

The demographics of 16 small countries in the Latin America and Caribbean region show that small countries may not be homogeneous in terms of the language, culture, or ethnicity of its people. For example, there are three dominant languages used in Suriname, a small country with a population of only 0.6 million. Cultural and language diversity creates extra demands on the resources and expertise required to develop/adapt curriculum, digital resources, and training materials for the implementation of ICT in the curriculum. Three proven strategies that could help policy makers in small developing states to address the various planning and implementation challenges are: (i) solicit and establish partnership arrangements with other organizations (e.g. private, international/regional and peer institutions) to plan and carry out ICT implementations; (ii) leverage international support, and (iii) establish multi-country collaborations.

References

Augustsson, N. (2014). Massive Open Online Courses (MOOCs): Challenging traditional learning. Foro Nacional Internacional. Agenda: Suramérica N°10. Lima, Peru. Retrieved from https://www.rockefellerfoundation.org/app/uploads/Agenda-Suramerica.pdf

Buabeng-Andoh, C. (2012). Factors influencing teachers' adoption and integration of information and communication technology into teaching: A review of the literature. *International Journal of Education and Development using Information and Communication Technology, 8*(1), 136–155.

Butcher, N., Moore, A., & Hoosen, S. (2014). Using UNESCO's ICT competency framework for teachers in Guyana. *Journal of Learning for Development-JL4D, 1*(2).

Central Intelligence Agency (CIA). (2015). *The world factbook.* Retrieved from https://www.cia.gov/library/publications/the-world-factbook/

Comisión Económica para América Latina y el Caribe (CEPAL). (1992). *Social policy in Costa Rica: a survey of recent reforms.* [La política social en Costa Rica: panorama de sus reformas recientes]. Retrieved from http://archivo.cepal.org/pdfs/1992/S9200654.pdf

Crossley, M., Bray, M., & Packer, S. (2009). Education in the small states of the Commonwealth: Towards and beyond global goals and targets. *The Round Table, 98*(405), 731–751.

Edmundson, A. L. (2011). The cultural adaptation of e-learning: Using appropriate content, instructional design, and media. In A. L. Edmunson (Ed.), *Cases on globalized and culturally appropriate e-learning: Challenges and solutions* (pp. 308–325). Hershey, PA: IGI Global.

Friedman, T. (2005). *The world is flat.* New York, NY: Farrar, Straus and Giroux.

Fallas, I., & Zúñiga, M. (2010). Las tecnologías digitales de la información y la comunicación en la educación costarricense. In *Tercer Informe de Estado de la Educación.* San José, Costa Rica: Programa Estado de la Nación. Retrieved from http://www.estadonacion.or.cr/files/biblioteca_virtual/educacion/003/Fallas_Zuniga_2010_TIC_Educacion.pdf

Gaible, E. (2009). *Survey of ICT and education in the Caribbean: A summary report based on 16 country surveys.* Washington, DC: infoDev/World Bank. Retrieved from http://www.infodev.org/infodev-files/resource/InfodevDocuments_441.pdf

Government of Canada. (2015). *Project profile: Strengthening distance education in the Caribbean.* Retrieved from http://www.acdi-cida.gc.ca/cidaweb%5Ccpo.nsf/projEn/A034764001

Guo, P. J., & Reinecke, K. (2014, March). Demographic differences in how students navigate through MOOCs. In *Proceedings of the first ACM conference on Learning @ scale conference - L@S '14* (pp. 21–30). New York, NY: ACM Press.

Henson, K. T. (2015). *Curriculum planning: Integrating multiculturalism, constructivism, and education reform* (5th ed.). Long Grove, IL: Waveland Press.

Hinostroza, J. E., & Labbé, C. (2011). *Policies and practices of educational computing in Latin America and the Caribbean* [Políticas y prácticas de informática educativa en América Latina y el Caribe]. Santiago, Chile: CEPAL.

Ho, A., Chuang, I., Reich, J., Coleman, C., Whitehill, J., Northcutt, C., … Petersen, R. (2015). HarvardX and MITx: Two years of open online courses [HarvardX Working Paper No. 10, 2015]. Cambridge, MA: Harvardx and MIT Office of Digital Learning.

Hood, N., Littlejohn, A., & Milligan, C. (2015). Context counts: How learners' contexts influence learning in a MOOC. *Computers & Education, 91*, 83–91.

Inter-American Development Bank. (2016). *IDBx Free online courses from Inter-American Development Bank*. Retrieved from https://www.edx.org/school/idbx

International Telecommunication Union - ITU-UN. (2015). *Percentage of individuals using the Internet* 2000-2014 [Data file]. Retrieved from http://www.itu.int/en/ITU-D/Statistics/Documents/statistics/2015/Individuals_Internet_2000-2014.xls

Kozma, R. B., & Vota, W. S. (2014). ICT in developing countries: Policies, implementation, and impact. In *Handbook of research on educational communications and technology* (pp. 885–894). New York, NY: Springer.

Law, N., Pelgrum, W. J., & Plomp, T. (Eds.). (2008). *Pedagogy and ICT use in schools around the world: Findings from the IEA SITES 2006 study*. Hong Kong: CERC, University of Hong Kong and Springer.

Lewis, M. W. (2011, March 26). Maps of Caribbean Geopolitics in *GeoCurrents* [Online image]. Retrieved from http://www.geocurrents.info/geopolitics/maps-of-caribbean-geopolitics

Lugo, M. T. (2010). Las políticas TIC en la educación de América Latina. Tendencias y experiencias. *Revista Fuentes, 10*, 52–68. Retrieved from https://ojs.publius.us.es/ojs/index.php/fuentes/article/view/2587/2415

Muñoz, L., Brenes, M., Bujanda, M. E., Mora, M., Núñez, O., & Zúñiga, M. (2014). *Las políticas TIC en los sistemas educativos de América Latina: Caso Costa Rica* (Vol. 4). Buenos Aires, Argentina: UNICEF.

Parrish, P., & Linder-VanBerschot, J. (2010). Cultural dimensions of learning: Addressing the challenges of multicultural instruction. *The International Review of Research in Open and Distributed Learning, 11*(2), 1–19.

Rogers, C. P., Graham, C. R., & Mayes, C. T. (2007). Cultural competence and instructional design: Exploration research into the delivery of online instruction cross-culturally. *Educational Technology Research & Development, 55*(2), 197–217.

Sachs, J. D., Mellinger, A. D., & Gallup, J. L. (2001). The geography of poverty and wealth. *Scientific American, 284*(3), 70–75.

Severin, E., & Capota, C. (2011). *One-to-one laptop programs in Latin America and the Caribbean: Panorama and perspectives. Panorama and perspectives*. Washington, DC: Inter-American Development Bank.

Sucre, F. & Garrett, M. B. (2016). *Educación basada en competencias en línea: Lecciones de EE.UU. para América Latina*. Inter-American Dialogue. Retrieved from http://www.thedialogue.org/wp-content/uploads/2016/03/Policy-Brief-EBCEL-FINAL.pdf

Suriname Ministry of Education and Community Development. (2014a). *National Policy and Strategy on ICT in Education – Suriname*. Retrieved from http://iicd.org/documents/national-policy-and-strategy-on-ict-in-education-report-of-current-status-of-ict-in-education/

Suriname Ministry of Education and Community Development. (2014b). *National strategic policy. ICT in education* [Nationaal Strategisch Beleidsplan ICT in het onderwijs]. Retrieved from http://www.freemanmc.com/images/easyblog_images/42/2ndBEIP-Nationaal-Strategisch-Beleidsplan-voor-ICT-in--het-onderwijs-2014.pdf

The Commonwealth. (2015). *The Nassau Declaration – 19th Conference of Commonwealth Education Ministers Education in the Commonwealth: Quality education for equitable development: Performance, paths and productivity (3Ps)*. Retrieved from http://www.adeanet.org/en/system/files/resources/nassau_declaration_22-26_june_2015_0.pdf

Trucano, M. (2012). *Quick guide: ICTs in education challenges and research questions*. Retrieved from http://www.infodev.org/articles/quick-guide-icts-education-challenges-and-research-questions

Trucano, M. (2014, March 18). *Two new rigorous evaluations of technology use in education* [Web log post]. Retrieved from http://blogs.worldbank.org/edutech/IDB-research

UNESCO. (2002). *Education for All: An international strategy to put the Dakar framework for action on Education for All into operation*. Paris, France: UNESCO.

UNESCO. (2012a). *ICT in education in Latin America and the Caribbean: A regional analysis of ICT integration and e-readiness*. Quebec: UNESCO.

UNESCO. (2012b). *Collaborative distance education in the Caribbean. An evaluation of the Caribbean Universities Project for Integrated Distance Education (CUPIDE)*. Kingston, Jamaica: UNESCO.

van der Hijden, P. (2014, July 27). *ICT in Education: Connect the schools, explore the world, develop Suriname* [Web log post]. Retrieved from https://sofosconsultancy.wordpress.com/2014/07/27/ict-in-education-connect-the-schools-explore-the-world-develop-suriname/

Wilke, E. (2010). *The influence of geography and physical ecology on economic development* (Unpublished Master's thesis). Southern Illinois University, Carbondale, IL. Retrieved from http://opensiuc.lib.siu.edu/theses/226

World Bank. (2011). *Preliminary draft: Master list of ICT/education policy documents (version 0.2)*. Retrieved from http://siteresources.worldbank.org/EDUCATION/Resources/WorldBankMasterICT-EdPolicyDocumentList-Draft_June2011_for-external-site.pdf

World Bank. (2015a). *Data indicators*. Retrieved from http://data.worldbank.org/indicator

World Bank. (2015b). *Country and lending groups*. Retrieved from http://data.worldbank.org/about/country-and-lending-groups#High_income

Author Information

Sdenka Zobeida Salas-Pilco
The University of Hong Kong
Faculty of Education, University of Hong Kong, Pokfulam Road, Hong Kong.

Complete mailing address: C/o Professor Nancy Law, Faculty of Education,
University of Hong Kong, Pokfulam Road, Hong Kong.
Telephone number: +852-62260287
Email address: sdenkasp@connect.hku.hk

Sdenka Zobeida Salas–Pilco earned her Ph.D. from the Faculty of Education at the University of Hong Kong. She obtained her Master in Educational Informatics (magna cum laude) from Alas Peruanas University, Peru. During her tenure as a secondary school teacher in the Peruvian Andes, she received scholarships to undertake specialized professional development on e-Learning in South Korea, on programming in India, and on social informatics in Japan. She has also worked in the Ministry of Education in Peru as an Information and Communication Technology (ICT) Specialist. Her Ph.D. research focussed on how ICT can be integrated into the curriculum of indigenous communities to stimulate interest, foster empowerment, and engender respect for local community knowledge. Her research interests include ICT policy in developing countries, learning designs, rural education, creative and informal environments, and sociocultural approaches to learning and development.

Nancy Law
The University of Hong Kong
Faculty of Education, University of Hong Kong, Pokfulam Road, Hong Kong.

Complete mailing address: Faculty of Education, University of Hong Kong, Pokfulam Road, Hong Kong.
Telephone number: +852-28592550
Fax number (optional): +852-25177194
Email address: nlaw@hku.hk
Website: http://web.edu.hku.hk/staff/academic/nlaw

Nancy Law was the Founding Director of the Centre for Information Technology in Education (CITE) at the University of Hong Kong and is currently serving as its Deputy Director. She is also the Corresponding Co-convenor for the Science of Learning Strategic Research Theme at the University. Professor Law is internationally known for her work in the area of e-learning and learning technologies, particularly in international comparative studies of pedagogical innovations using IT, models of ICT integration in schools and change leadership, and IT-supported knowledge building for students, teachers and professional communities. A unique feature of her research is the depth of analyses and the remarkable breadth spanning many levels, from classrooms and schools to whole educational systems—characteristics she considers to be essential to address, both in theory and in practice, the sustainability and scalability of IT-supported educational innovations. Her current research focus is to apply design-based research methods to build network models of innovation that integrate teacher professional development and school leadership development with pedagogy, assessment, and learning technology co-design. She is developing a technology platform, the Learning Design Studio, to support the changing role of teachers from instruction professionals to designers of learning experiences.

Chapter 5
New Challenges for ICT in Education Policies in Developing Countries: The Need to Account for the Widespread Use of ICT for Teaching and Learning Outside the School

J. Enrique Hinostroza

Abstract The widespread availability and use of ICT in society, in general, is leading teachers and students to use these tools to complement their teaching and learning processes outside the traditional classroom context. However, the evidence shows that many students and teachers lack the digital skills needed to make effective use of these tools; which limits their potential impact, can have negative consequences for students' learning, and can increase educational inequalities, especially in developing countries. Against this backdrop, based on a literature review of international studies and ICT in educational policies of developing countries, this chapter presents and discusses the challenges to ICT in education policies that this new scenario poses, including the definition of the set of digital skills that teachers and students need to acquire, the integration of these skills in the curriculum, the design and implementation of strategies to ensure that teachers and students develop these skills, and the design and implementation of pedagogical methods that take advantage of these skills in the teaching and learning of traditional subject areas. Based on this evidence, the chapter highlights the challenges and opportunities to address.

Keywords ICT in Education Policies in Developing Countries • Use of ICT outside the School • Second digital divide

J. Enrique Hinostroza (✉)
Instituto de Informática Educativa, Universidad de La Frontera,
Montevideo, 0830 Temuco, Chile
e-mail: enrique.hinostroza@iie.cl

I.A. Lubin (ed.), *ICT-Supported Innovations in Small Countries and Developing Regions*, Educational Communications and Technology: Issues and Innovations,
DOI 10.1007/978-3-319-67657-9_5

1 Introduction

Since the early 1980s, there have been country-level efforts to introduce and use computers in schools. As Cuban (2001) argues, justifications for these efforts have been associated with three different aims: making schools more efficient and productive, transforming teaching and learning into an active process connected to real life, and preparing the current generation of young people for the future workplace.

Considering these aims during the last three decades, which is what the chapter describes as the old scenario, the implementation of Information and Communication Technology in education (ICTE) policies in many developed and developing countries has progressed in three phases: providing infrastructure, developing support mechanisms for the use of ICT, and aligning ICTE policy with a broader educational vision and set of policies.

The goal of phase 1 has been to start with the provision of infrastructure, (productivity) software, and basic ICT skills training, in order to ensure access to ICT. This initial phase of policy implementation has been in many cases accompanied by government policies aimed at the infusion of ICT in society and/or industry. For example, Korea defined its first Master Plan for ICT in Education in 1996, with the main goal of establishing the required ICT infrastructure in the country (Keris, 2014). Similarly, Rwanda adopted the One laptop per Child Initiative in 2007, to set-up up the technical infrastructure and observe children's interactions with the laptops.[1] In a similar vein, during the late 1990s and early 2000s, many countries in Latin America and the Caribbean defined their ICTE policies and all of them included objectives related to the provision of computers and Internet connection to schools and some type of teacher training strategy (Hinostroza & Labbé, 2011).

Following this initial stage, countries have defined policies that show an emphasis on creating support mechanisms for the use of ICT, including for example, technical and pedagogical support, putting special attention on the use of ICT in teaching and learning. Thus, in many cases, phase 2 included the implementation of national support structures to ensure the adequate conditions and incentives for the use of ICT in the classrooms. For example, in Chile, the ICTE policy – Enlaces, created a Technical Assistance Network, based on a strategic alliance between the Ministry of Education and 24 universities across the country. The network's mission was to train teachers and provide them with technical and educational support (Hinostroza, Hepp, & Cox, 2009). Likewise, the *Plan Ceibal* initiative in Uruguay provided one laptop to each student in the country, and after the initial round of infrastructure provision, started the implementation of a strategy to support teachers in the use of the laptops, first using a cascade model and then a peer teacher training strategy (Hinostroza, Jara, & Brun, 2011).

Also of consequence during the last few years, there has been an emphasis on the development of teachers' ICT skills and competencies, as well as an emerging trend

[1] See: http://wiki.laptop.org/images/5/53/Rwanda_Report-v7.pdf

to use mobile and/or home based ICT infrastructure as tools to extend teaching and learning possibilities. This is the case despite the differences in the level of ICT development across countries. For example, the UNESCO ICT Competency Framework for Teachers (UNESCO, 2012) has been widely promoted across countries.

In addition, many of the phase 2 policies also show an emphasis that goes beyond the school settings, including the use of ICT in the community and an emphasis on a broader variety of hardware and software (both stand-alone and on-line). For example, the already mentioned *Plan Ceibal* has been both a social and an educational policy that has the purpose of contributing to social equity. Thus, it aims to reduce the digital gap in the country; promote social justice by establishing the conditions for equal access to ICT; facilitate the construction of new learning environments adequate to demands of the information and knowledge-based society; and make available to students and teachers new tools that can widen their learning, increase their knowledge, and develop their awareness of lifelong learning (Hinostroza et al., 2011).

Finally, the third phase of implementation of ICTE policies, particularly in developing countries, has been characterized by a stronger commitment to aligning ICTE policies with a broader educational vision as well as to define more precise objectives associated with the development of students' digital skills. This is driven by the vision of developing a 'smart society' in the country, such as in the case of Korea.

Across these phases, ICT has been associated with educational change and innovation (Fullan, 2012), considering it as a catalyst that will trigger the change (McDonald & Ingvarson, 1997) and as a Trojan Horse (Olson, 2000) that brings about innovation. Such views of technology have been especially attractive to policy designers, especially in developing countries who associate its incorporation to the modernization of the school system. However, as it is argued in this chapter, the current situation shows that ICT is not only under-delivering such promises, but also, due to the rapid availability of access and use of ICT in and out of school, the increased unsupervised use, and the lack of teachers' and students' digital competencies, developing countries are in danger of undermining quality and equality of education.

1.1 Current Status of ICT in Education: Persistent Challenges and New Possibilities

While ICT in education policies of the past three decades were intended to have a positive impact on the educational landscape, new questions are being raised about what these impacts are and how precisely they are being realized. Veritably, some old educational challenges have persisted even as researchers and educators see new possibilities for the future.

First, despite the promises explicitly or implicitly included in the ICTE in education policies that are related to the positive impact of ICTs on students' achievements, there is no conclusive evidence to support these promises (OECD, 2015), especially in developing countries (Hinostroza, Isaacs, & Bougroum, 2014). This situation has posed new questions to the research community and policy makers, who are now looking for much more precise definitions of the role of ICT in teaching and learning.

Second, the use of ICT in classroom settings is still limited in frequency and variety (Fraillon, Ainley, Schulz, Friedman, & Gebhardt, 2014; Law, Pelgrum, & Plomp, 2008). Even so, in settings outside the school, both students and teachers have more access to ICT and use it much more frequently and creatively for teaching and learning purposes (Hinostroza, Ibieta, Claro, & Labbé, 2016; Meneses, Fàbregues, Rodríguez-Gómez, & Ion, 2012; OECD, 2015; Wang, Hsu, Campbell, Coster, & Longhurst, 2014). This is true for many developing countries. This situation is offering new opportunities for researchers and policy makers, who are looking on with great interest. An example is the perspective of Learning Lives that investigates learning opportunities beyond the traditional school context (Erstad, 2012).

Third, a persistent challenge is in identifying and defining the digital skills that are essential for an effective participation in the information society (van Dijk & van Deursen, 2014). This challenge applies to learning in general (Calvani, Fini, Ranieri, & Picci, 2012; Fraillon et al., 2014) and in relation to specific learning aims, such as mathematics and reading (OECD, 2015). This problem is being embraced by a number of research projects aiming at the definition of these digital skills (see for example: Ananiadou & Claro, 2009; Binkley et al., 2012), how to effectively develop them (see for example: Azevedo, Johnson, Chauncey, & Burkett, 2010; Brand-Gruwel, Wopereis, & Walraven, 2009; Kinley, Tjondronegoro, Partridge, & Edwards, 2014), as well as factors affecting their development (see for example: Hatlevik & Christophersen, 2013; Jara et al., 2015).

These conditions form part of a new scenario that is very much framed by the development and use of digital technologies in society. The next section in this chapter examines this new scenario, presenting evidence showing that despite the widespread access to ICT, teachers' and students' capacity to harness its potential in education is still limited. Against this backdrop, drawing on a review of ICT in education policies from some developing countries[2] and the results of international studies, the third section presents the implications for ICTE policies; teacher training and development; and curriculum design, development, and implementation. Based on this evidence, the final section summarizes the challenges and opportunities that need to be addressed, especially in small developing countries.

[2] Due to the authors' research background and expertise, the examples are mainly from Latin America and especially from Chile.

2 The New Scenario

The increasing availability and access to Information and Communication Technologies (ICT), especially in developing countries (ITU, 2014), is changing multiple dimensions of the ways in which people learn, work, spend their free time, communicate, and engage with others and themselves. These changes have an impact on how societies are organized, tending to an increased virtualization of culture and the development of horizontal interactive communication networks (Castells, 2004). Furthermore, as ICT have expanded in society, they are influencing home and family life, changing dynamics, lifestyles, and cultural expectations (Facer, 2012; Facer, Furlong, Furlong, & Sutherland, 2003; Livingstone & Helsper, 2007).

In particular, Internet access has increased significantly among young people, who use it from different locations (home, school) and for different types of activities (Ito, 2010; Livingstone, Haddon, Görzig, & Ólafsson, 2011; Pedró, 2012; Rideout, Foehr, & Roberts, 2010). For example, as shown in Fig. 5.1, in the Latin American countries that participated in PISA[3] 2009 and 2012, the access to Internet at home increased by around 20%. Students in Uruguay reported having the best access, representing an increase from 60% to 82%. On the opposite end of the spectrum, Peruvian students reported having the least access to Internet at home, but still recorded a 17% increase from 25% to 42%.

Indeed, a significant number of young people have grown up with digital technology, the Internet, video games, and mobile telephones; technologies that are being used for formal and informal learning, as well as social and recreational purposes (Erstad, 2012; Facer et al., 2003; Furlong & Davies, 2012; Hinostroza, Matamala, Labbé, Claro, & Cabello, 2014; Livingstone & Helsper, 2007). Moreover, teachers are also increasingly using ICT to prepare materials and communicate with students outside the traditional classroom boundaries (Hinostroza et al., 2016; Meneses et al., 2012; Wang et al., 2014).

In this emerging 'virtualized' context, educational systems are compelled to effectively contribute to the development of citizens with new skills and abilities, often referred to as twenty-first century skills (Ananiadou & Claro, 2009; Binkley et al., 2012; Fraillon, Schulz, & Ainley, 2013). This demand presents a series of challenges to policy makers, schools principals, teachers, students, and families. Among others, these challenges include transforming the curriculum, assessments, pedagogical methods, and learning styles (Dede, 2010; Kozma, 2011a).

From a policy perspective, educational systems are acknowledging these challenges. In fact, results of the 2013 International Computer and Information Literacy Study (ICILS) showed:

> Nearly all national centers [of the participating countries] identified the following as important aspects of educational policies and plans: preparing students to use ICT as a learning tool, development of information literacy, and development of ICT-based skills in critical thinking, collaboration, and communication. (Fraillon et al., 2014, p. 56)

[3] Programme for International Student Assessment – https://www.oecd.org/pisa/home/

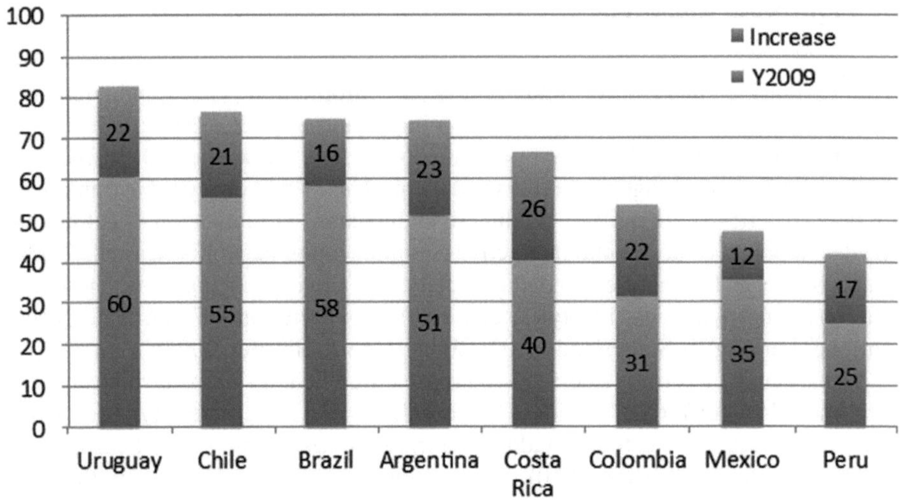

Fig. 5.1 Percentage of students who reported having access to the Internet at home in 2009 and the increase in the percentage in 2012. OECD (2015)

However, results of ICILS also show that 40% of the participating students perform at level one or below and 78% perform at level two or below (see the definition of the levels in Table 5.1).

This implies that almost three-quarters of the students have only sufficient skills to use computers to complete basic and explicit information-gathering and information-management tasks. In particular, these students are able to locate explicit information from within given electronic sources; make basic edits and add content to existing information products in response to specific instructions; create simple information products that show consistency of design and adherence to layout conventions; demonstrate awareness of mechanisms for protecting personal information; and demonstrate awareness of some of the consequences of public access to personal information. Conversely, only 2% of the students are able to do the following: select the most relevant information to use for communicative purposes; evaluate usefulness of information based on criteria associated with need and evaluate the reliability of information based on its content and probable origin; create information products that demonstrate a consideration of audience and communicative purpose; use appropriate software features to restructure and present information in a manner that is consistent with presentation conventions; and adapt information to suit the needs of an audience (Fraillon et al., 2014).

Aligned with this state of affairs, Fig. 5.2 shows the results of the digital reading test of PISA 2012. In particular, the results indicate that 37%, 29%, and 55% of students in Brazil, Chile, and Colombia respectively, perform under level 2. This implies that "they have difficulties using conventional navigation tools and features, and locating links or information that are not prominently placed" (OECD, 2015, p. 91). These results, when compared with the 18% average for the OECD countries participating in PISA 2012, portray a rather challenging scenario.

Table 5.1 Summary of the levels of the CIL scale

Level	Definition
Level 4 (above 661 scale points)	Students working at Level 4 select the most relevant information to use for communicative purposes. They evaluate the usefulness of information based on criteria associated with need and evaluate the reliability of information based on its content and probable origin. These students create information products that demonstrate a consideration of audience and communicative purpose. They also use appropriate software features to restructure and present information in a manner that is consistent with presentation conventions and they adapt that information to suit the needs of an audience. Students working at Level 4 also demonstrate awareness of problems that can arise with respect to the use of proprietary information on the Internet.
Level 3 (577 to 661 scale points)	Students working at Level 3 demonstrate the capacity to work independently when using computers as information-gathering and information-management tools. These students select the most appropriate information source to meet a specified purpose, retrieve information from given electronic sources to answer concrete questions, and follow instructions to use conventionally recognized software commands to edit, add content to, and reformat information products. They recognize that the credibility of web-based information can be influenced by the identity, expertise, and motives of the creators of that information.
Level 2 (492 to 576 score points)	Students working at Level 2 use computers to complete basic and explicit information-gathering and information-management tasks. They locate explicit information from within given electronic sources. These students make basic edits and add content to existing information products in response to specific instructions. They create simple information products that show a consistency of design and adherence to layout conventions. Students working at Level 2 demonstrate awareness of mechanisms for protecting personal information. They also demonstrate awareness of some of the consequences of public access to personal information.
Level 1 (407 to 491 score points)	Students working at Level 1 demonstrate a functional working knowledge of computers as tools and a basic understanding of the consequences of computers being accessed by multiple users. They apply conventional software commands to perform basic communication tasks and add simple content to information products. They demonstrate familiarity with the basic layout conventions of electronic documents

Source: Fraillon et al. (2014, pp. 19–20)

Overall, the results are consistent with other studies showing that many students lack the skills for more advanced uses of Internet, for example, for solving information problems (van Deursen & van Diepen, 2013), challenging thereby the concept of 'digital natives' (Thompson, 2013). Moreover, these studies show that students' more advanced ICT skills are largely dependent on their economic, social, and cultural capital. Therefore, some groups appear to gain more benefits for ICT, thus increasing social inequalities (Brandtzæg, Heim, & Karahasanović, 2011; Hargittai & Hinnant, 2008; Jara et al., 2015; OECD, 2015; van Dijk & van Deursen, 2014). These findings can easily be extrapolated to country level scenarios, where small developing countries are clearly in disadvantageous positions compared to more developed countries.

In summary, the new scenario proposed here is characterized by a widespread availability and use of ICT by students and teachers, yet a relatively low capacity to harness its potentials. This new scenario has several consequences for the design and implementation of ICT in education policies, which are presented and discussed in the next section.

3 Implications for Educational Policy and Teacher and Curriculum Development

The scenario described in the previous section bears several implications for ICT in education policies; teacher training and development; and curriculum design, development, and implementation that are described in this section. The main argument here is that this new scenario is challenging traditional designs of education policies and their components, highlighting the need to change the traditional question from how to use ICT to transform education, to a new one that asks how to teach and learn in a transformed digital context.

3.1 Educational Policies

In relation to educational policies, there are at least four implications that need attention. First, despite the promises and intentions included in the ICTE policies, the purposeful integration of ICT in classroom activities is still a challenge; more so is the implementation of innovative pedagogies empowered by ICT as a widespread practice in schools. Second, regardless of the investment in equipping schools with computers and the Internet, the integration of ICT in teaching and learning activities is happening much more often outside the classroom context, however, very few countries, if any, can ensure that teachers have the required understanding of this situation and that they possess the knowledge and competencies to manage students' learning outside the classroom. Third, also as a consequence of the ever-increasing availability of educational resources on the Internet, the practice of

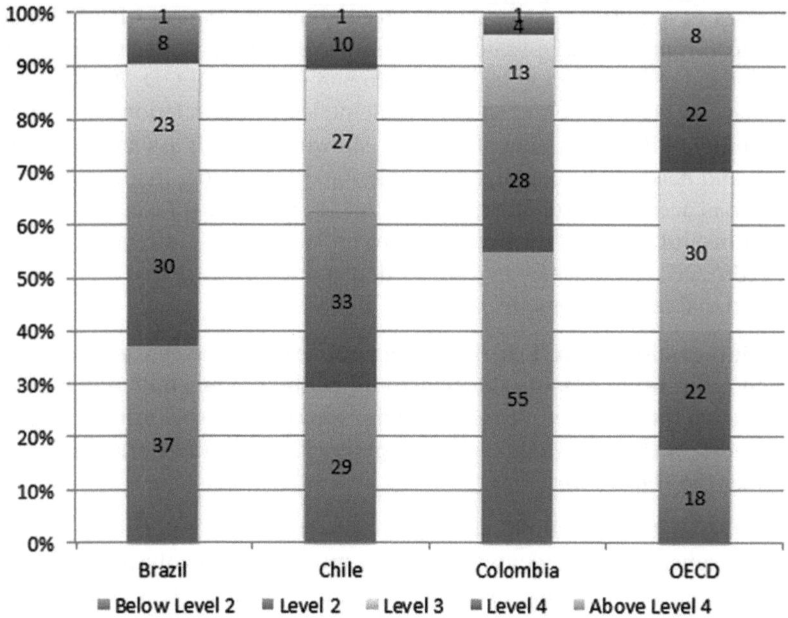

Fig. 5.2 Results in PISA 2012 Digital Reading Test of selected countries

copying and pasting information directly from the Internet for schoolwork is spreading rapidly among students, highlighting the need to ensure that teachers and students are aware of the ethical and legal aspects of using information available on the Internet. Fourth, the above call for redesigning the type of learning activities that are often incorporated in the national curriculum or school textbooks, so as to ensure that they take pedagogical advantage of the new digital context. Below these implications are discussed in further detail.

1. **The purposeful integration of ICT in classroom activities is still a challenge; more so is the implementation of innovative pedagogies empowered by ICT as a widespread practice in schools.**

As mentioned before, many ICT in education policies have proposed the use of digital tools as a catalyst (McDonald & Ingvarson, 1997), a lever (Venezky, 2002), or a Trojan Horse (Olson, 2000) for change and innovation in the classroom that, in turn, would improve students learning achievements. However, results of large-scale studies show that the reality in schools lags considerably behind these and other promises of technology (OECD, 2015). In fact, although there is a growing body of evidence associating specific uses of particular digital tools in teaching and learning to positive impacts on students' outcomes (Tamim, Bernard, Borokhovski, Abrami, & Schmid, 2011; Underwood, 2009), in general, these results correspond to small-scale interventions that are difficult to scale-up and have been created in developed countries contexts or under quite particular conditions elsewhere

(Hinostroza et al., 2014). Also, the evidence about the impact of ICT on student performance found in large-scale interventions is at best inconclusive, contradictory, or mixed (Hinostroza et al., 2014; Kozma, 2011a; OECD, 2015). Empirical studies have not been able to consistently demonstrate the relation between the use or availability of ICTs and student learning, showing that this association is far more complex and needs to be explained considering other factors that could lie beyond mere use and access (Claro, 2010; Kozma, 2011b; Pedró, 2012). In addition, studies show that use of digital tools in the classroom is relatively infrequent (Law et al., 2008; Pedró, 2011) and when they are used, it is often the case that they merely support traditional teaching practices (Kozma, 2003).

2. **The integration of ICT in teaching and learning activities is happening much more often outside the classroom context. The challenge is to ensure that teachers have the required understanding of this situation and that they possess the knowledge and competencies to manage students' learning outside the classroom.**

Reports of national and international studies show that it is relatively commonplace for teachers to ask students to complete projects that involve using the Internet to practice content, searching for information, and using the text processor to draft reports and presentations (Hogarty, Lang, & Kromrey, 2003; Hsu, 2011; OECD, 2013; Tondeur, Van Braak, & Valcke, 2007). Coherently, studies show that students report using computers and the Internet at home for schoolwork quite regularly (Hinostroza, Matamala, Labbé, Claro, & Cabello, 2014; OECD, 2015).

Although encouraged by teachers, these activities are not always supervised or guided by them. In fact, it is quite common that teachers do not define the type of information sources that students should use, the number of sources required, how to evaluate the quality and relevance of the information found, etc. (van Deursen & van Diepen, 2013). Therefore, students tend to use their intuition to search and select information instead of the formal procedures or methods, such as the Information Problem Solving while using the Internet, proposed by Brand-Gruwel et al. (2009). Moreover, studies have shown that when students use a Web search engine, they tend toward a "get in, get the answer, get out" approach, rather than using strategies that could foster deeper learning on the web (Thompson, 2013, p. 21). This approach may threaten the potential benefits of using the Internet for teaching and learning since students could potentially be using partial or misleading sources of information.

3. **There is a need to ensure that teachers and students are aware of the ethical and legal aspects of using information available on the Internet.**

The third and fourth challenges can be considered a consequence of the ever-increasing availability of educational resources on the Internet. That is, on the one hand, the Internet contains a wealth of resources that are being used for teaching and learning, which is generally assumed as one of the most important contributions. This is especially salient in small developing countries, where access to up-to-date information and learning resources is in many cases difficult or impossible. On the other hand, the Internet is also a wealth of ready-made answers that students can easily copy and paste in order to complete their school assignments. In fact, plagiarism and Internet dishonesty have been largely studied in higher education and are now becoming relevant issues in K-12 settings. Furthermore, some studies show that plagiarism is a widespread behavior among secondary students in European countries (Dias & Bastos, 2014).

This growing problem is seen as an unfortunate corollary of the increase of new technologies and easy access to information (Dias & Bastos, 2014). Among the reasons for this phenomenon, Evering and Moorman (2012, p. 38) argue that "when Millennials are given vague or busywork assignments, their lack of interest may prompt them to plagiarize just to get the task done, or their lack of understanding of how to do the assignment may cause them to turn to plagiarism to maintain their grade point average" (p. 38). Overall, evidence suggests that the habit of copy and pasting ready-made answers from the Internet is a growing phenomenon that requires attention (Pedró, 2012). This is especially the case considering the relative lack of research that specifically studies this phenomenon in schools.

4. **There is a need to review and eventually redesign the type of learning activities that are often incorporated in the national curriculum or school textbooks, so as to ensure that they take into consideration (are aware of) the new digital context.**

In addition to the ethical considerations already discussed, there is also the problem that students complete their school assignments without engaging in the learning and cognitive processes that the activities are designed to trigger (Evering & Moorman, 2012). For example, in many developing countries there is a relatively long tradition (in teacher education) of designing learning activities based on Bloom's taxonomy (Bloom, Englehart, Furst, Hill, & Krathwohl, 1956; Krathwohl, 2002). This taxonomy suggests that specific types of activities can be used to trigger certain cognitive processes (learning objectives). However, if students simply copy and paste the answer from the Internet, they will not engage in the expected activities and the expected cognitive processes will not be triggered.

Importantly, and especially in developing countries, it is relatively common that Ministries of Education provide free textbooks for students attending public schools, which include guidelines suggesting classroom activities for each lesson. For example, in Chile, studies[4] have shown that more than 90% of teachers in Chile use the textbooks provided by the Ministry of Education in more than half of their lessons. They have also shown that textbooks are mainly used to carry out activities with students, including assignments. As a consequence, it is easy to find the answers to the assignments on the Internet, especially considering that many of them are the same from one year to the other.

[4] The study was performed by Centro de Microdatos of the Universidad de Chile for the Ministry of Education in 2013. Available at: http://www.textosescolares.cl/index2.php?id_seccion=3753&id_portal=65&id_contenido=15838

3.2 *Teacher Training and Development*

For at least two decades, ICT in education policies have incorporated and implemented strategies to enable teachers to use ICT in their classrooms and elsewhere. Despite these efforts, results of TALIS 2013[5] show that "on average, the second and third most important professional development needs teachers report are related to teaching with ICT skills (19% of teachers) and to using new technologies in the workplace (18% of teachers)" (OECD, 2014, p. 109).

This situation is consistent with research showing that teachers lack the required digital skills to make effective use of these tools in the classroom, and in relation to the emerging digital skills described previously, they don't know strategies to teach such skills to their students (van Deursen & van Diepen, 2013). However, studies show that teachers are using ICT relatively frequently outside the classroom and that most the frequent activities are those related to lesson preparation, such as creating resources and using the Internet to search for and gather material (Hogarty et al., 2003; Hsu, 2011; Tondeur et al., 2007), further elaborate content, searching for classroom resources, and using the text processor to elaborate materials (Hsu, 2011; Russell, Bebell, O'Dwyer, & O'Connor, 2003).

Considering this scenario, the following implications emerge from the new context, specifically associated with teachers.

1. **The need to define the set of digital competencies that teachers need to (a) make effective use of ICT in their professional activities and (b) teach these skills to their students**.

Although there are comprehensive definitions of teachers' ICT competencies, such as the UNESCO (2011) ICT Competency Framework for Teachers, considering that the definition of these skills is still a matter of discussion, how to teach them is still under discussion as well (Krumsvik, 2012). For example, the OECD study, ICT and Initial Teacher Education, found that "Teacher competencies are not always well-defined and, even when they are, it is often the case that they do not endorse a clear vision of what teaching and learning in a knowledge society should be and what supporting role technology can play" (Rizza, 2011, p. 40). This implies that new teachers can be not only underprepared to effectively use ICT in their teaching practices, but also, they will not necessarily be aware of how to develop students' digital skills.

2. **The need to ensure that new teachers are equipped with the required digital competencies**.

In relation to the integration of ICT in initial teacher training institutions, some studies have shown that the integration of ICT is limited to a few specific resources (mainly computers and projectors), mostly applied to perform 'traditional'

[5] The OECD Teaching and Learning International Survey – http://www.oecd.org/edu/school/talis.htm

pedagogical activities (Brun & Hinostroza, 2014; Rizza, 2011). This is probably a major challenge that requires sound policy level definitions to provide appropriate incentives for these institutions to advance in the integration of ICT in their curriculum and practice.

3. **The need to define and implement appropriate incentives for teachers to integrate ICT in their professional activities, including using ICT during classroom activities.**

In relation to educational policy, many educational systems incorporate standards and incentives for teachers that define the expected professional performance. However, very few of them incorporate the use of ICT in teaching. For example, in an OECD (2013) review, teachers' digital skills are (almost) not mentioned, showing that the majority of the educational systems considered in the report are not integrating them in their formal system level instruments.

3.3 Curriculum and Students

The growing access to digital technology has led some authors to label newer generations as 'digital natives' (Prensky, 2001; Prensky & Berry, 2001). These digital natives are credited as possessing a series of skills that are required in the digital world (Thompson, 2013). However, empirical evidence has prompted several authors to question this concept (Hargittai, 2010; Kirschner & Karpinski, 2010; Margaryan, Littlejohn, & Vojt, 2011; Ng, 2012; Waycott, Bennett, Kennedy, Dalgarno, & Gray, 2010). In particular, recent studies suggest that the range of technologies used by young people is much more restricted than is commonly assumed. Their use of technology is often superficial and when it comes to searching for information they do so hastily and without following a more elaborate, iterative strategy (Thompson, 2013). The reasons that are suggested for this, include a lack of skills for working with information on the Internet in a more structured way (Kirschner & Karpinski, 2010), traditional learning styles, and little technological fluency (Margaryan et al., 2011). They also show that, despite popular belief, students do not have the ability to carry out several activities simultaneously. Instead, they process information sequentially, with constant changes of context that negatively affect the learning outcome (APA, 2006). Considering that there is not enough evidence confirming that new generations have the innate capacity to take full advantage of ICT, the following implications emerge concerning the curriculum.

1. **The need to formally introduce aims associated with the development of digital/ICT skills in the curriculum, including definitions and corresponding operationalization.**

Regarding the type of digital skills to incorporate into the curriculum, although definitions are still under discussion, there is a relative consensus that they include computer and information literacy, which are defined by the ICILS as "an individual's ability to use computers to investigate, create, and communicate in order to participate effectively at home, at school, in the workplace, and in society" (Fraillon et al., 2013, p. 17). Aligned with this, some countries are starting to define their own frameworks to emphasize the need to develop these skills. For example, the Chilean Ministry of Education defined digital skills as "the capacity to solve information, communication and knowledge problems as well as legal, social and ethical dilemmas in digital contexts"(Mineduc, 2013, p. 17). Based on this definition, they defined a set of competencies grouped in for areas: Information, Communication, Digital Coexistence, and ICT. In a similar vein, as shown in Table 5.2 (adapted from Ferrari, 2013), the Joint Research Centre of the European Commission defined digital competence as the "confident, critical and creative use of ICT to achieve goals related to work, employability, learning, leisure, inclusion and/or participation in society" (p. 2).

2. **The need to purposefully integrate these skills into traditional subject areas.**

In this regard, considering that the results of digital reading skills of PISA 2012 showed that "there is a strong association between digital Reading and the quality of navigation (task-oriented browsing), even after accounting for performance in print Reading" (OECD, 2015, p. 121). However, on average, only 25% of the students in the participating countries showed highly focused browsing activity, while 61% showed insufficient or mixed browsing activity, and 12% mostly unfocused browsing activity or no browsing activity (3%). In this sense, it seems reasonable, for example, to discuss the incorporation of navigation skills in the Language curriculum, in order to prepare students to use digital sources of information more effectively.

3. **The need to develop focalized programs aimed at leveling students' access and use of ICT, especially in relation to the skills needed to effectively take advantage of the Internet, students' beliefs about the resources, and possibly students' Internet use habits**.

From a different perspective, and directly affecting students, there is a growing body of evidence showing that, once access to technology is leveled, students' ability to make full use of these technologies is very much associated with their socioeconomic background (Hargittai, 2008, 2010), academic performance (van Deursen & van Diepen, 2013), academic self-efficacy (Greene, Yu, & Copeland, 2014), epistemic beliefs (Strømsø & Bråten, 2010), self-regulated learning strategies (Lazonder & Rouet, 2008), and a set of cognitive and behavioural characteristics that shape the quality of how they use ICT (Jara et al., 2015).

The variability in students' capacity to take advantage of ICT, has lead researchers and international organizations to warn about the emergence of a second digital divide (OECD, 2010), which does not only impact students' use of ICT itself, but also the possibility of some having available significantly more and better resources to learn from and with. Therefore, without proper policies, there is a risk of exacerbating existing educational differences, particularly those related to students' socio-economic backgrounds. This is especially relevant in small developing countries where educational systems tend to show high levels of inequality (Cox, 2010).

Table 5.2 Areas of digital competence defined by the Joint Research Centre of the European Commission

Area	Definition
Information	Identify, locate, retrieve, store, organize and analyze digital information, judging its relevance and purpose
Communication	Communicate in digital environments, share resources through online tools, link with others and collaborate through digital tools, interact with and participate in communities and networks, cross-cultural awareness.
Content-creation	Create and edit new content (from word processing to images and video); integrate and re-elaborate previous knowledge and content; produce creative expressions, media outputs, and programming; deal with and apply intellectual property rights and licenses.
Safety	Personal protection, data protection, digital identity protection, security measures, safe and sustainable use.
Problem-solving	Identify digital needs and resources, make informed decisions on most appropriate digital tools according to the purpose or need, solve conceptual problems through digital means, creatively use technologies, solve technical problems, update own and other's competence.

Regarding their beliefs about the availability of resources and possibly their children's Internet use habits, incorporating guidelines and suggestions for parents is also recommended. In particular, research shows that many parents believe that simply because of the access to ICT, students are motivated to learn and improve their academic performance, which is not the case (Hayward, Alty, Pearson, & Martin, 2002). Therefore, guidelines should seek to reinforce parents' awareness and their strategies to mediate children's use of ICT, including monitoring, control, and counseling strategies (Yu, Yuen, & Park, 2012).

4 Concluding Remarks

In the previous sections it has been argued that due to the widespread availability and use of ICT in society, teachers and students are using these tools outside the traditional classroom context to complement their teaching and learning processes. However, results of international studies and research show that many students and teachers lack the digital skills needed to make effective use of these tools, which limits their potential impact, can have negative consequences for students' learning, and can increase educational inequalities, especially in developing countries. This new scenario poses challenges to ICT in education policies that need to be addressed, including the definition of the set of digital skills that teachers and students need to acquire, the integration of these skills in the curriculum, the design and implementation of strategies to ensure that teachers and students develop these skills, and the design and implementation of pedagogical methods that take advantage of these skills in the teaching and learning of traditional subject areas.

Regarding the design of ICT in educational policies, in their study of meta-analyses examining educational interventions that improved learning outcomes in low- and middle-income countries, Evans and Popova (2015) found broad support across the reviews for "(i) pedagogical interventions that match teaching to students' learning, including through the use of computers or technology; (ii) individualized, long-term teacher training; and (iii) accountability-boosting interventions, such as teacher performance incentives and contract teachers" (Evans & Popova, 2015, p. 3).

These findings can be associated with the recommendations made in the previous section. Regarding the need to match teaching to students' learning, it seems clear that the way in which students are taught in their classrooms does not match what and how they learn in other contexts. Therefore, several challenges mentioned before address the need to align teaching and learning strategies, as well as the curriculum, taking into consideration the availability and use of digital contents and tools.

In relation to teachers, as mentioned previously, there seems to be a consensus that despite the large investment in teacher training initiatives, there is a piece missing that prevents teachers from using ICT in their regular lessons. In this sense, training teachers to ensure that they possess the required digital skills so as to take advantage of the full potential of ICT for their professional activities can be considered a first step towards facilitating their appropriation of these tools.

Finally, regarding the incentives, recommendations made refer to the definition and incorporation of ICT standards and incentives for teachers' use of ICT in their professional activities, especially, incorporating these competencies in national teacher appraisal systems that are linked to their career progression.

Although these recommendations can be applied to ICT in Education policies in general, the rapid availability of access and use of ICT in developing countries, the high degree of inequality of educational systems in developing contexts, and the increasing unsupervised (intuitive) use of ICT to complement teaching and learning activities outside of school, make these recommendations especially relevant for developing countries, including the very smallest. Thus, considering that in many cases governments in developing countries associate the provision of ICT to a modernization process of the educational system, but without preparing teachers and students to effectively take advantage of these tools, these efforts can lead to an illusion of modernity that in fact is increasing inequalities due to inadequate use of ICT.

Acknowledgements This revision was possible thanks to the contribution from National Fund for Scientific and Technological Development (Fondecyt No. 1151044).

References

Ananiadou, K., & Claro, M. (2009). *21st century skills and competencies for new millennium learners in OECD countries* [OECD Education Working Papers Series]. Paris, France: Organisation for Economic Co-operation and Development.

APA. (2006). *Multitasking: Switching costs*. Retrieved from http://www.apa.org/research/action/multitask.aspx

Azevedo, R., Johnson, A., Chauncey, A., & Burkett, C. (2010). Self-regulated learning with MetaTutor: Advancing the science of learning with MetaCognitive tools. In M. S. Khine & I. M. Saleh (Eds.), *New science of learning: Computers, cognition and collaboration in education* (pp. 225–247). New York, NY: Springer.

Binkley, M., Erstad, O., Herman, J., Raizen, S., Ripley, M., Miller-Ricci, M., & Rumble, M. (2012). Defining twenty-first century skills. In P. Griffin, B. McGaw, & E. Care (Eds.), *Assessment and teaching of 21st century skills* (pp. 17–66). New York, NY: Springer Science+Business Media B.V.

Bloom, B., Englehart, M., Furst, E., Hill, W., & Krathwohl, D. (1956). Taxonomy of educational objectives: The classification of educational goals. In *Handbook I: Cognitive domain.* New York, NY: Longmans, Green.

Brand-Gruwel, S., Wopereis, I., & Walraven, A. (2009). A descriptive model of information problem solving while using Internet. *Computers & Education, 53*(4), 1207–1217. https://doi.org/10.1016/j.compedu.2009.06.004

Brandtzæg, P. B., Heim, J., & Karahasanović, A. (2011). Understanding the new digital divide—A typology of Internet users in Europe. *International Journal of Human-Computer Studies, 69*(3), 123–138. https://doi.org/10.1016/j.ijhcs.2010.11.004

Brun, M., & Hinostroza, J. E. (2014). Learning to become a teacher in the 21 st century: ICT integration in initial teacher education in Chile. *Educational Technology & Society, 17*(3), 222–238.

Calvani, A., Fini, A., Ranieri, M., & Picci, P. (2012). Are young generations in secondary school digitally competent? A study on Italian teenagers. *Computers & Education, 58*(2), 797–807. https://doi.org/10.1016/j.compedu.2011.10.004

Castells, M. (Ed.). (2004). *The network society: A cross-cultural perspective.* Northampton, MA: Edward Elgar.

Claro, M. (2010). *Impacto de las TIC en los aprendizajes de los estudiantes.* Estado del arte *CEPAL,* Colección documentos de proyecto Santiago de Chile.

Cox, C. (2010). Educational inequality in Latin America. Patterns, policies and issues. In P. Attewell & K. S. Newman (Eds.), *Growing gaps: Educational inequality around the world* (pp. 33–58). New York, NY: Oxford University Press.

Cuban, L. (2001). *Oversold and underused.* London, England: Harvard University Press.

Dede, C. (2010). Technological supports for acquiring 21st century skills. *International Encyclopedia of Education, 3*, 158–166.

Dias, P. C., & Bastos, A. S. C. (2014). Plagiarism phenomenon in European countries: Results from GENIUS Project. *Procedia - Social and Behavioral Sciences, 116*, 2526–2531. https://doi.org/10.1016/j.sbspro.2014.01.605

Erstad, O. (2012). The learning lives of digital youth—Beyond the formal and informal. *Oxford Review of Education, 38*(1), 25–43. https://doi.org/10.1080/03054985.2011.577940

Evans, D. K., & Popova, A. (2015). *What really works to improve learning in developing countries? An analysis of divergent findings in systematic reviews* [Policy Research Working Paper]. Washington, DC: World Bank.

Evering, L. C., & Moorman, G. (2012). Rethinking plagiarism in the digital age. *Journal of Adolescent & Adult Literacy, 56*(1), 35–44. https://doi.org/10.1002/jaal.00100

Facer, K. (2012). Taking the 21st century seriously: Young people, education and socio-technical futures. *Oxford Review of Education, 38*(1), 97–113. https://doi.org/10.1080/03054985.2011.577951

Facer, K., Furlong, J., Furlong, R., & Sutherland, R. (2003). *Screenplay: Children and computing in the home.* London, England: RoutledgeFalmer.

Ferrari, A. (2013). *DIGCOMP: A Framework for developing and understanding digital competence in Europe* [Scientific and Policy Report]. Seville, Spain: Joint Research Centre of the European Commission.

Fraillon, J., Ainley, J., Schulz, W., Friedman, T., & Gebhardt, E. (2014). *Preparing for life in a digital age: The IEA International Computer and Information Literacy Study International Report.* Springer Open.

Fraillon, J., Schulz, W., & Ainley, J. (2013). *International computer and information literacy Study: Assessment framework*. Amsterdam, The Netherlands: International Association for the Evaluation of Educational Achievement (IEA).

Fullan, M. (2012). *Stratosphere: Integrating technology, pedagogy, and change knowledge*. Toronto, ON: Pearson Canada.

Furlong, J., & Davies, C. (2012). Young people, new technologies and learning at home: Taking context seriously. *Oxford Review of Education, 38*(1), 45–62. https://doi.org/10.1080/03054985.2011.577944

Greene, J. A., Yu, S. B., & Copeland, D. Z. (2014). Measuring critical components of digital literacy and their relationships with learning. *Computers & Education, 76*, 55–69. https://doi.org/10.1016/j.compedu.2014.03.008

Hargittai, E. (2008). The digital reproduction of inequality. In D. Grusky (Ed.), *Social stratification*. Boulder, CO: Westview Press.

Hargittai, E. (2010). Digital na(t)ives? Variation in Internet skills and uses among members of the "Net Generation". *Sociological Inquiry, 80*(1), 92–113. https://doi.org/10.1111/j.1475-682X.2009.00317.x

Hargittai, E., & Hinnant, A. (2008). Digital inequality. Differences in young adults' use of the Internet. *Communication Research, 35*(5), 602–621.

Hatlevik, O. E., & Christophersen, K.-A. (2013). Digital competence at the beginning of upper secondary school: Identifying factors explaining digital inclusion. *Computers & Education, 63*, 240–247. https://doi.org/10.1016/j.compedu.2012.11.015

Hayward, B., Alty, C., Pearson, S., & Martin, C. (2002). *Young people and ICT 2002: Findings from a survey conducted in autumn 2002*. London, England: British Telecommunications and Technology Agency (BECTA).

Hinostroza, J. E., Hepp, P., & Cox, C. (2009). National policies and practices on ICT in education: Chile (Enlaces). In T. Plomp, R. E. Anderson, N. Law, & A. Quale (Eds.), *Cross-national information and communication technology: Policies and practices in education* (revised second edition) (pp. 153–170). Greenwich, England: Information Age Publishing.

Hinostroza, J. E., Ibieta, A., Claro, M., & Labbé, C. (2016). Characterisation of teachers' use of computers and Internet inside and outside the classroom: The need to focus on the quality. *Education and Information Technologies, 21*(6), 1595–1610. https://doi.org/10.1007/s10639-015-9404-6

Hinostroza, J. E., Isaacs, S., & Bougroum, M. (2014). Information and communications technologies for improving students' learning opportunities and outcomes in developing countries. In D. Wagner (Ed.), *Learning and education in developing countries: Research and policy for the Post-2015 UN Development Goals* (pp. 42–57). New York, NY: Palgrave Macmillan.

Hinostroza, J. E., Jara, I., & Brun, M. (2011). Chapter 6: Case study: Uruguay. In R. B. Kozma (Ed.), *Transforming education: The power of ICT policies* (pp. 125–164). Paris, France: UNESCO.

Hinostroza, J. E., & Labbé, C. (2011). *Policies and practices for the use of information and communications technologies (ICTs) in education in Latin America and the Caribbean Políticas Sociales*. Santiago, Chile: Economic Commission for Latin American and the Caribbean (ECLAC).

Hinostroza, J. E., Matamala, C., Labbé, C., Claro, M., & Cabello, T. (2014). Factors (not) affecting what students do with computers and Internet at home. *Learning, Media and Technology, 40*(1), 43–63. https://doi.org/10.1080/17439884.2014.883407

Hogarty, K. Y., Lang, T. R., & Kromrey, J. D. (2003). Another look at technology use in classrooms: The development and validation of an instrument to measure teachers' perceptions. *Educational and Psychological Measurement, 63*(1), 139–162.

Hsu, S. (2011). Who assigns the most ICT activities? Examining the relationship between teacher and student usage. *Computers & Education, 56*(3), 847–855. https://doi.org/10.1016/j.compedu.2010.10.026

Ito, M. (2010). *Hanging out, messing around, and geeking out: Kids living and learning with new media*. Cambridge, MA: MIT press.

ITU. (2014). *Measuring the information society*. Geneva, Switzerland: Author.
Jara, I., Claro, M., Hinostroza, J. E., San Martín, E., Rodríguez, P., Cabello, T., ... Labbé, C. (2015). Understanding factors related to Chilean students' digital skills: A mixed methods analysis. *Computers & Education, 88*, 387–398. https://doi.org/10.1016/j.compedu.2015.07.016
KERIS. (2014). *White Paper on ICT in Education Korea: Korean Ministry of Education (MOE) and Korea Education and Research Information Service (KERIS).*
Kinley, K., Tjondronegoro, D., Partridge, H., & Edwards, S. (2014). Modeling users' web search behavior and their cognitive styles. *Journal of the Association for Information Science and Technology, 65*(6), 1107–1123. https://doi.org/10.1002/asi.23053
Kirschner, P. A., & Karpinski, A. C. (2010). Facebook® and academic performance. *Computers in Human Behavior, 26*(6), 1237–1245. https://doi.org/10.1016/j.chb.2010.03.024
Kozma, R. B. (Ed.). (2003). *Technology, innovation and educational change: A global perspective*. Eugene, OR: International Society for Technology in Education - International Association for the Evaluation of Educational Achievement.
Kozma, R. B. (2011a). The technological, economic, and social contexts for educational ICT policy. In R. B. Kozma (Ed.), *Transforming education: The power of ICT policies* (pp. 3–18). Paris, France: UNESCO.
Kozma, R. B. (2011b). *Transforming education: The power of ICT policies*. Paris, France: Unesco.
Krathwohl, D. R. (2002). A revision of Bloom's taxonomy: An overview. *Theory Into Practice, 41*(4), 212–218.
Krumsvik, R. J. (2012). Teacher educators' digital competence. *Scandinavian Journal of Educational Research, 58*(3), 269–280. https://doi.org/10.1080/00313831.2012.726273
Law, N., Pelgrum, W. J., & Plomp, T. (2008). *Pedagogy and ICT use in schools around the world: Findings from the IEA SITES 2006 study*. Hong Kong: Springer Comparative Education Research Centre.
Lazonder, A. W., & Rouet, J.-F. (2008). Information problem solving instruction: Some cognitive and metacognitive issues. *Computers in Human Behavior, 24*(3), 753–765. https://doi.org/10.1016/j.chb.2007.01.025
Livingstone, S., Haddon, L., Görzig, A., & Ólafsson, K. (2011). *Risks and safety on the Internet. The perspective of European children. Full findings (pp. 56)*. London: EU and LSE.
Livingstone, S., & Helsper, E. J. (2007). Gradations in digital inclusion: Children, young people and the digital divide. *New Media & Society, 9*(4), 671–696. https://doi.org/10.1177/1461444807080335
Margaryan, A., Littlejohn, A., & Vojt, G. (2011). Are digital natives a myth or reality? University students' use of digital technologies. *Computers & Education, 56*(2), 429–440. https://doi.org/10.1016/j.compedu.2010.09.004
McDonald, H., & Ingvarson, L. (1997). Technology: A catalyst for educational change. *Journal of Curriculum Studies, 29*(5), 513–527.
Meneses, J., Fàbregues, S., Rodríguez-Gómez, D., & Ion, G. (2012). Internet in teachers' professional practice outside the classroom: Examining supportive and management uses in primary and secondary schools. *Computers & Education, 59*(3), 915–924. https://doi.org/10.1016/j.compedu.2012.04.011
Mineduc. (2013). *Matriz de habilidades TIC para el aprendizaje*. Santiago, Chile: Ministerio de Educación – Centro de Educación y Tecnología – ENLACES.
Ng, W. (2012). Can we teach digital natives digital literacy? *Computers & Education, 59*(3), 1065–1078. https://doi.org/10.1016/j.compedu.2012.04.016
OECD. (2010). *Are the New Millennium learners making their grade? Technology use and educational performance in PISA*. Paris, France: Organisation for Economic Co-operation and Development – Center for Educational Research and Innovation.
OECD. (2013). *Teachers for the 21st century: Using evaluation to improve teaching*. Paris, France: Author.
OECD. (2014). *TALIS 2013 results: An international perspective on teaching and learning*. Paris, France: Author.

OECD. (2015). *Students, computers and learning: Making the connection*. Paris, France: Author.

Olson, J. (2000). Trojan Horse or teacher's pet? Computer and the culture of the school. *Journal of Curriculum Studies, 32*(1), 1–8.

Pedró, F. (2011). *Tecnología en la escuela: Lo que funciona y porqué*. Madrid, Spain: Fundación Santillana.

Pedró, F. (2012). *Connected minds: Technology and today's learners*. Paris, France: OECD Publishing.

Prensky, M. (2001). Digital natives, digital immigrants part 1. *On the Horizon, 9*(5), 1–6.

Prensky, M., & Berry, B. D. (2001). Do they really think differently? *On the Horizon, 9*(6), 1–9.

Rideout, V. J., Foehr, U. G., & Roberts, D. F. (2010). *Generation M2: Media in the lives of 8- to 18-year-olds* [A Kaiser Family Foundation Study]. Menlo Park, CA: The Henry J. Kaiser Family Foundation.

Rizza, C. (2011). *ICT and initial teacher education: National policies* [OECD Directorate for Education Working Paper]. Paris, France: Organisation for Economic Co-operation and Development.

Russell, M., Bebell, D., O'Dwyer, L., & O'Connor, K. (2003). Examining teacher technology use implications for preservice and inservice teacher preparation. *Journal of Teacher Education, 54*(4), 297–310.

Strømsø, H. I., & Bråten, I. (2010). The role of personal epistemology in the self-regulation of Internet-based learning. *Metacognition and Learning, 5*(1), 91–111. https://doi.org/10.1007/s11409-009-9043-7

Tamim, R. M., Bernard, R. M., Borokhovski, E., Abrami, P. C., & Schmid, R. F. (2011). What forty years of research says about the impact of technology on learning: A second-order meta-analysis and validation study. *Review of Educational Research, 81*(1), 4–28.

Thompson, P. (2013). The digital natives as learners: Technology use patterns and approaches to learning. *Computers & Education, 65*(July), 12–33. https://doi.org/10.1016/j.compedu.2012.12.022

Tondeur, J., Van Braak, J., & Valcke, M. (2007). Towards a typology of computer use in primary education. *Journal of Computer Assisted Learning, 23*(3), 197–206.

Underwood, J. (2009). *The impact of digital technology: A review of the evidence of the impact of digital technologies on formal education*. Coventry, England: British Educational Communications and Technology Agency (Becta).

UNESCO. (2011). *UNESCO ICT competency framework for teachers*. Paris, France: Author.

UNESCO. (2012). Turning on mobile learning in Asia: Illustrative initiatives and policy implications. In F. Pedró (Ed.), *UNESCO working paper series on mobile learning*. Paris, France: Author.

van Deursen, A. J., & van Diepen, S. (2013). Information and strategic Internet skills of secondary students: A performance test. *Computers & Education, 63*, 218–226. https://doi.org/10.1016/j.compedu.2012.12.007

van Dijk, J. A. G. M., & van Deursen, A. J. A. M. (2014). *Digital skills, unlocking the information society*. Basingstoke, England: Palgrave Macmillan.

Venezky, R. L. (2002, March 6). *Quo Vademus? The transformation of schooling in a networked world*. Retrieved from http://www.oecd.org/dataoecd/48/20/2073054.pdf

Wang, S.-K., Hsu, H.-Y., Campbell, T., Coster, D. C., & Longhurst, M. (2014). An investigation of middle school science teachers and students use of technology inside and outside of classrooms: Considering whether digital natives are more technology savvy than their teachers. *Educational Technology Research and Development, 62*(6), 637–662. https://doi.org/10.1007/s11423-014-9355-4

Waycott, J., Bennett, S., Kennedy, G., Dalgarno, B., & Gray, K. (2010). Digital divides? Student and staff perceptions of information and communication technologies. *Computers & Education, 54*(4), 1202–1211. https://doi.org/10.1016/j.compedu.2009.11.006

Yu, M., Yuen, A. H. K., & Park, J. (2012). Students' computer use at home: A study on family environment and parental influence. *Research and Practice in Technology Enhanced Learning, 7*(1), 3–23.

Author Information

J. Enrique Hinostroza
Instituto de Informática Educativa, Universidad de La Frontera, Montevideo 0830, Temuco, Chile

Complete mailing address: Instituto de Informática Educativa, Universidad de La Frontera, Montevideo 0830, Temuco, Chile
Telephone number: +56 45 2325252
Email address: enrique.hinostroza@iie.cl

J. Enrique Hinostroza is the Director of the Institute of ICT in Education of the Universidad de La Frontera in Temuco, Chile. During the last decade, he has acted as a senior consultant for the Chilean ICT in Education Program Enlaces as well as for other countries in Latin America and the Caribbean region, and for international agencies and companies, including UNESCO, ECLAC, World Bank, Inter-American Development Bank, Brookings and Intel CAG. He has conducted several national and international research projects and published in a number of international journals. He has professional qualifications in Industrial Engineering (1998), a Masters degree in Computer Science (1990) and a Ph.D. in Information Technology in Education (1999 – Institute of Education, University of London). His research interests are related to the understanding and development of digital skills, the use of ICT in teaching and learning and the design and evaluation of ICT in education policies. Current research projects grants include the development of digital skills for learning through learning activities mediated by digital assistants and the identification of the characteristics of school assignments that promote higher order cognitive processes and better search strategies of secondary students.

Chapter 6
Implementing Sustainable ICT-Supported Innovation Policies: Case of Universitas Terbuka – Indonesia

Dewi A. Padmo and Tian Belawati

Abstract As a distance learning university, the use of ICT at Universitas Terbuka (UT) is of central importance for both administration and instruction. As technology continues to advance and change very rapidly, the university is pressured to enhance the utilization of ICT even more. It is believed that the use of an ICT-supported distance learning environment can and will increase the quality of student learning as well as enhance the efficiency of the administration. Yet, two important considerations are the nascent national ICT infrastructure and the lack of readiness of students to embrace ICT-supported learning environments. Given the challenging and limited national ICT infrastructure, the challenge for management is how to keep innovating and introducing state-of-the-art ICT-supported learning to facilitate the enhancement of student readiness and their familiarity with online learning. Similarly, strong leadership is needed to convince the entire university faculty and staff to support the university's policies on the development and implementation of ICT–supported innovations. This chapter will describe the experience of Universitas Terbuka (The Indonesia Open University) in addressing concerns about the sustainability of their ICT-Supported Learning System and administration, including discussions about policy-making, action plans, implementation, student participation, student and staff ownership, as well as maintenance.

Keywords Policies in implementing ICT • Sustainable ICT • Open universities • ICT learning systems and administration

---

The original version of this chapter was revised. An erratum to this chapter can be found at DOI 10.1007/978-3-319-67657-9_9

D.A. Padmo (✉) • T. Belawati
Universitas Terbuka,
Jl. Cabe Raya, Pondok Cabe, Pamulang, South Tangerang 15418, Banten, Indonesia
e-mail: dewi@ecampus.ut.ac.id; http://www.ut.ac.id/; tbelawati@ecampus.ut.ac.id; http://www.ut.ac.id/

I.A. Lubin (ed.), *ICT-Supported Innovations in Small Countries and Developing Regions*, Educational Communications and Technology: Issues and Innovations, DOI 10.1007/978-3-319-67657-9_6

1 Background

Indonesia is the largest archipelago country in the world with over 17,000 islands located in the South-East Asia region. In terms of population, Indonesia is the fourth largest nation in the world with a population of over 255 million people (CIA Factbook, 2016). Amid abundant natural resources, Indonesia's human development index is still classified as medium and ranks at 110 in the 2014 Human Development Index, with Gross National Income per capita of US$ 9787 (UNDP, 2015; World Bank, 2016). Nevertheless, the McKinsey Global Institute (2016) predict that Indonesia can become the 7th largest economy in the world by 2030 if it can increase its base of skilled workers from about 55 to 135 million. This is based on the fact that Indonesia's dependency ratio – a measure of the percentage of the population too young or too old to work – is about 40. This means that Indonesia is currently enjoying the demographic benefit of a higher productive-age group within the population. However, if Indonesia fails to upgrade the competency and capacity of its people, this demographic bonus will not benefit the country, but instead can become a burden to the economy. Therefore, education has become the top priority for the government, such that in a 2002 amendment to the constitution, the government allocated 20% of the national budget to the educational sector (Republic of Indonesia, 2003).

Indonesia's formal education system consists of four levels. There are the primary, junior secondary, and senior secondary levels with participation rates of about 96%, 74%, and 54% respectively. Then there is the tertiary level which, despite a large number of higher education institutions in Indonesia (about 4200), has a much lower participation rate – at about 32% (UNDP, 2015). This low participation rate is due to several barriers, including the geographical situation, as well as the unequal spread of higher education institutions within the 37 provinces in the country (see Fig. 6.1). As seen in the map, Indonesia is an archipelago, which consists of six big islands (Sumatera, Java, Kalimantan, Sulawesi, Nusa Tenggara, Maluku, and Papua) and thousands of small and remote islands that are underdeveloped in terms of infrastructure. Most higher education institutions can be found on the islands of Java and Sumatera, which are located in the southwest part of the country. The uneven distribution of educational infrastructure, of course, is far from ideal and is jeopardizing the realization of Indonesia becoming a large-economy country.

The advancement of information and communication technologies (ICT) opens up many possibilities for both broadening access to and increasing quality of education (Kozma & Vota, 2014). Today, it is not only that education can be provided online, but also that there are many educational resources available on the Internet. The global movement on openness has resulted in the development of ample open source software, open content, and even open courses that can be beneficial for the development of human capital (Belawati, 2014). It is now possible for any country to enrich its provision of education by taking advantage of what is available in the digital world. Yet, for developing countries, there are numerous impediments to obtaining what value there is in open education through ICT (Hinostroza & Labbé, 2011).

Fig. 6.1 Map of Indonesia (Source: Indahnesia.com, 2017)

For example, as data shows, most of the available open content is in languages foreign to Indonesia, such as English, Spanish, Chinese, French, etc. (Daryono & Belawati, 2013). Another requirement for realizing the benefits of open educational content is the availability of Internet access (Daryono & Belawati, 2013).

For Indonesia, content in local languages and access to the Internet remain issues that need to be addressed. Although English is taught in school from Grade 3 of primary education level, most Indonesians are not fluent in either speaking or reading English. Most Indonesian people speak at least two languages, including their local language and the national language of Bahasa Indonesia. Additionally, with regards to Internet access, although Indonesia ranks number four in Asia, the penetration rate is still only around 34%, which is much lower than Malaysia (68%), and Singapore (81%) (Internet Worldstats, 2016).

The Government of Indonesia (GoI) has responded by trying to increase access to higher education. Understanding the wide geographical context of the country, the GoI realized that it was not possible to build enough campuses to equitably satisfy the population demand. The government's Ministry Education and Culture, following the United Kingdom Open University (UKOU) model, decided in 1984 to establish an open university, which later was named the Indonesian Open University - better known as Universitas Terbuka (UT). The main purpose of UT's establishment was to increase the capacity of the higher education system in providing access to all citizens. Therefore, UT was designed to deliver quality distance learning programs using the most appropriate technology, so that the university would be able to tap students' unreached potential. Those are the people who, due to various constraints including lack of funding, living in isolated rural areas, and lack of time are not able to continue their studies at university level. Thus, the reasons for the establishment of UT were both educational as well as political in nature. UT represents the presence of the government in providing access to education to all citizens, as mandated by the 1945 Indonesian Constitution (Republic of Indonesia, 2016).

2 Policies for ICT Development and Implementation

The instructional system of UT followed that of UKOU, which was based on pre-produced learning materials, tutorial services, and examinations. In the year of UT's establishment, it was decided that UT's main learning medium would be printed learning materials supplemented by various audio and video materials. This policy implied that 100% of curriculum content must be conveyed in forms of printed materials known as Modules, and any audio and video materials were meant as enrichment and supplements only. In line with the level of technology at the time, learning support services were mainly provided in forms of face-to-face tutorials and limited television and radio broadcasts. Likewise, examinations were also designed as pencil and paper tests conducted in appointed locations in each region. The decision to provide self-contained and self-instructional learning materials was taken considering that many potential students might not have any access to other media other than the printed books.

Nevertheless, as a distance learning university with a large number of students, UT has always relied on technology to support its data management and administration. In the beginning, UT's computing system was developed chiefly for administrative purposes, one of which is the main application for the Student Record System (SRS). The SRS stores and manages data on curriculum, student registrations, as well as examinations and grades (Belawati, Hardhono, Sinar, & Patria, 2009). It is this SRS that later became the backbone of the entire university computing system. Since the introduction of the Internet in Indonesia in the mid-1990s, UT immediately developed the university website which later became the basis for the current comprehensive UT Online Learning System.

To ensure sustainability, policies on any development, including the development of various ICT applications, are comprehensively stated in the university's strategic and operational plans. Therefore, the commitment to become a more modern open university by optimizing the use of ICT is also formulated in UT's strategic plan. The current 2010–2021 Strategic Plan (Universitas Terbuka, 2011a) emphasizes the importance and need to improve the quality of administration and instruction through the use of appropriate ICT. The Strategic Plan, which is further elaborated in the 2010–2013 and the 2014–2017 Operational Plans, lays out the annual stages of development and targets to be achieved. Each target is well defined in measurable output indicators and the relevant Vice Rector (Vice President), through respective working units within the university, monitors the progress. Although it is led by the Computer Center, the responsibilities of each stage of ICT development are shared and involve many people and units working to increase the sense of ownership of the output among people and units. This is very important to ensure the seamless process of implementing the ICT applications once they are deployed. To ensure the sustainability of implementations, all policies are well disseminated and actions are all consistently taken through a comprehensive process of development, training, implementation, maintenance, and evaluation based on the outlined targets and indicators of achievements.

The basic infrastructure for the ICT system at UT was established in the years prior to the current Strategic Plan. The infrastructure includes a Private Virtual Network (PVN) that electronically connects the Head Office in Jakarta with all 39 Regional Offices throughout Indonesia. Having this PVN is considered very strategic to enhance the unification/integration of data, to increase the speed and accuracy of data exchanges, as well as to increase management efficiency. Therefore, in the 2010–2021 Strategic Plan, ICT development is more focused on the utilization and alignment of all supporting applications for both administration and academic purposes.

ICT *utilization* refers to the enhancement of the use of ICT for all of the university's affairs including registration, tuition payment and finance, development and distribution of learning materials, provision of learning supports, management and dissemination of information, management and organization of examinations, as well as various administrative applications (Sinar, 2014). On the other hand, ICT *alignment* refers to harmonization of all ICT applications to support all activities comprehensively to all users (Sinar, 2014). In 2008, to assist the planning of the enhancement of ICT utilization, UT developed an ICT grand design, which later was integrated into the university's Strategic Plan for 2010–2021. This grand design was updated in 2013. In general, the ICT development in the 2010–2021 Strategic Plan can be categorized into two main streams, which are administration/operational and academic/instructional affairs.

3 ICT-Supported Administration System

As an open university catering to a large number of students, UT has always relied on technology to support its administration affairs. As mentioned previously, the first computer application that was developed and heavily used by UT was the Student Record System (SRS), which has become the backbone of the whole university computing system. A further step in the development of the ICT-supported administration system was the establishment of a Private Virtual Network (PVN) that electronically connected all UT offices (Patria, 2014). All other development to support administrative works is well outlined in UT's Strategic Plan 2010–2021, which is further elaborated in the Operational Plan of 2010–2013 and the Operational Plan of 2014–2017 (Universitas Terbuka, 2014a).

3.1 Policy and Action Plans on ICT-Supported Administration System

The Strategic Plan of 2010–2021 has set several targets to be achieved with regard to the ICT-supported administration system. The priorities of ICT development in the Operational Plans of 2010–2013 and 2014–2017 for administration and operational affairs include the improvement of the online registration system that is integrated with the payment system, the establishment of a contact center, and the improvement of human resource management that is integrated with the financial system.

As outlined in the Operational Plans, the specific targets for the above development are as follows. At the end of 2017:

- The online registration system will be well established and used by at least 3% of students
- The ICT-based financial management system will be running efficiently and will be used for budget planning, implementing, and reporting
- The ICT-based asset management system will be running efficiently and will be used at both university and unit levels
- The contact center will be well established, and
- The ICT-based human resource management system that is integrated with the remuneration system will be running efficiently and used for payroll

3.2 Implementation of the ICT-Supported Administration System

UT began developing online registration in 2011 and gradually implemented it in 2012 (Milwan, 2014). The system is designed to help enrolled students as well as prospective students to register easily by simply accessing the website, http://sro.ut.ac.id. Table 6.1 shows the use of the online registration system in the years 2012–2015 for both new students (first registration) and for students who were already enrolled (course registration). Compared to the total registration (approximately 400,000 students), Table 6.1 shows that the use of online registration is still very low. This low number is understandable since students are not yet familiar with an online registration system. However, increases in the number of students using the online registration system, especially in the Diploma/Bachelor program, indicate that the innovation is appreciated.

The application system for financial management is one of the most crucial ICT developments. The financial management applications that have been developed and used are those for planning, budget execution, and reporting. ICT applications that have been used can be divided into two types: the unit-level for generating the internal financial planning to be integrated at the university level, and the university-level for financial planning that is directly connected with the Ministry of Finance of Indonesia (Damayanti, 2014). Similarly, for the realization of the budget, UT has been using applications that are connected directly to the Ministry of Finance. The use of these applications is helpful for financial management, from planning to reporting. As for asset management, UT has also been using ICT applications both for procurement of goods and services, as well as for recording all

Table 6.1 Online registration use

	Diploma/Bachelor program		Master program	
Semester	First registration	Course registration	First registration	Course registration
2012.1	336	4910	–	–
2012.2	472	5687	–	–
2013.1	67	1849	193	1413
2013.2	456	3960	388	1361
2014.1	117	3789	336	923
2014.2	498	4638	699	997
2015.1	349	5608	153	1388
2015.2	544	6.169	693	1180

Source: Universitas Terbuka (2012, 2013, 2014b, 2015)

university assets. Furthermore, since 2013, procurement of goods and services has also been done using e-procurement applications.

A Contact Center, an ICT innovation for opening up a multi-platform communication channel with students, was launched in 2014. Known as 'Hallo UT', the center manages and responds to all queries about UT received either through telephone calls, e-mails, faxes, SMS, or social media (Facebook, Twitter, LinkedIn, and Google plus). Through 'Hallo UT', students or the general public can access information about UT, convey any complaints or concerns, and ask questions related to the UT services. In addition, UT's Contact Center has also become an agent for Digital Marketing.

Regarding the ICT-supported human resource management, UT developed SIMPEG, which is an application for handling personnel administration that can be accessed through the website, http://simpeg.ut.ac.id. Through this application, employees' data are stored for easy retrieval. The system is additionally utilized for processing employee promotion, evaluation of performance and productivity, as well as the payroll. Through this application, appraisal of each employee's performance is processed automatically to calculate the amount of remuneration for each employee. Thus, the human resource application is well-integrated with the financial applications.

4 ICT-Supported Learning System

With a large number of students to be served, UT cannot provide a timely quality learning system without the use of ICT. In fact, this is stated as university policy in UT's Strategic Plan as well as in the Operational Plans. The main goals of providing an ICT-supported learning system are to address the need for more accessible learning materials, online learning supports, and examinations. Under these umbrellas, the targets include, but are not restricted to, the following: easy access to UT academic products through UT's Digital Library, availability of online laboratories in the forms of dry labs for practical work, availability of online self-tests, increased access to UT's learning supports through the online platform and the short text messages (SMS) platform, as well as access to a reliable online examination system.

4.1 Policy and Action Plans on ICT-Supported Learning System

As mentioned earlier, UT's Strategic Plan Year 2010–2021 is further divided into two action plans, namely the Operational Plan of 2010–2013 and 2014–2017, respectively (Universitas Terbuka, 2011a, 2014a). Both operational plans specifically define the objective and targets of their respective ICT developmental programs and activities.

The specific targets outlined in the Operational Plan of 2010–2013 and 2014–2017 related to the ICT-supported learning system include the following, which are expected by the end of 2017:

- All printed course materials will have been digitized and can be accessed through the Digital Library's Virtual Reading Room
- All practical work courses will be available in forms of dry labs in the university learning portal
- All courses will be provided with online tutorial services
- UT will offer online examinations for all courses in both multiple-choice and essay formats
- UT will offer at least three fully-online post-graduate programs including the use of tablet-based learning materials
- UT will offer massive open online courses (MOOCs) for at least one course per faculty as parts of its community services
- All online learning supports will be accessible through a variety of communication devices, including mobile devices

It is envisaged that at the end of 2017, UT students will have the benefit of choices of learning systems. As a result of the continuous development and improvement of the ICT-supported learning system, students will be able to choose whether they would like to pursue their education at UT through the online platform or the more conventional distance learning platform (i.e. printed based materials with a face-to-face tutorial system). Students will also be able to access various learning supports and resources through both online and offline systems at the Regional Offices. Importantly, maintaining the offline system is very important to ensure the inclusiveness of the UT system to people who do not have access to the Internet. Based on the Internet world statistics, the number of Internet users in Indonesia is about 34% (88 million out of 255 million of the total population). Thus, about 66% of the population does not have access to the Internet (Internet Worldstats, 2016).

4.2 Implementation of ICT-Supported Learning System

The implementation of the ICT-supported learning system, in accordance with the Operational Plans, has been conducted purposively and systematically. As intended, a range of innovative uses of ICT in improving the quality of the learning system is now available for students. All innovations are meant to help students improve their independent learning processes, and therefore the benefit of such innovations is measured through the students' utilization of the products and services.

At the end of 2014, UT had digitized learning materials for about 985 courses, which are made available to students through the Virtual Reading Room (Universitas Terbuka, 2014b). The existence of the Virtual Reading Room is well appreciated by the students. Data show that in its first year of deployment in 2013, the number of visitors to the Virtual Reading Room reached 152,000, 62.5% of whom were unique visitors or repeat visitors (Wahyono, 2014). This shows that UT's policy on ICT-supported learning materials distribution (digitized learning materials distributed through the Digital Library) has reached almost 30% of its total students.

The online tutorial provision is at the core of the ICT-supported learning support system at UT. As targeted, at the end of 2013 UT had provided online tutorial services for all courses using the Moodle-based Learning Management System (LMS) (Moodle, 2016). The improvement of the online tutorial services is not only in terms of the number, but also in the quality (Zuhairi, Adnan, & Thaib, 2007). Online tutors are trained and required to enrich their tutorial materials with visualization of concepts, including those taken from open educational resources (OERs) available from various Internet sources. By adding OER materials to the online tutorials, students are likely to be more motivated and interested in their learning, as well as to expand their horizons. The use of OER by academic staff relates to an effort in providing pedagogical supports for learners, such as "exciting learners about new ways of learning, providing learners with current knowledge, and supporting learner

Table 6.2 Total number of online tutorials offered and the number of participants Year 2011–2015

Semester	Number of tutorials offered	Number of tutorials with participants	Total number of participants in courses
2011.1	664	649	23,504
2011.2	660	628	31,578
2012.1	747	698	64,118
2012.2	752	719	105,078
2013.1	930	849	108,942
2013.2	930	843	123,960
2014.1	1021	985	222,905
2014.2	1044	1013	226,479
2015.1	1063	1045	277,397
2015.2	1076	1070	319,235

Source: Universitas Terbuka (2011b, 2012, 2013, 2014b, 2015)

creativity" (Dhanarajan & Abeywardena, 2013, p. 11). To illustrate the impact of policy on online tutorials, Table 6.2 shows the increase in the number of online tutorials and their participants from 2011 to 2015.

As seen in Table 6.2, the number of online tutorial courses offered had increased steadily. Similarly, the number of online tutorials with participants attending and the number of overall enrolled participants has also increased significantly. In the second Semester of 2015, the number of students who participated in online tutorials was 319,235 – an increase of 1,258% within the duration of 4 years. Participants in the second semester of 2015 were participating within the online tutorials of 1070 courses and were divided into around 4000 virtual classrooms, with each virtual classroom attended by 35–112 student participants.

In addition, as part of the online learning support system, UT also developed an online self-test system as an exercise platform for students, especially in preparation for the final examination. Similar to the online tutorials, this service is also well appreciated by the students as shown by the steady increase in utilization as presented in Table 6.3. The table shows the number of online self-tests accessed by students and the total number of students accessing the online self-tests. Moreover, to help students with practical work, all courses needing such work have been supplemented with dry labs that are integrated within the online tutorials. A dry lab is a computer-based simulation of experiments that helps students to become familiar with conceptual, logistical, and numerical aspects of a practical experience before coming to the laboratory (Learmonth, 1996). These efforts demonstrate the consistency of UT policies in optimizing the use of ICT in all aspects of learning services.

The steady increase in the utilization of both online tutorial and online self-test services suggests that the policy on ICT-supported learning system is well valued by the students. Students' Internet accessibility has improved as well, as reported by Internet Worldstat (2016). The Internet penetration in Indonesia in 2016 was about 34% (88 million out of the total of 250 million population). While still small, it represents an increase of about 1.0% from 2015 (Internet Worldstat, 2016). The increased accessibility to the Internet has also been influenced by mobile device penetration. Data show that mobile phone or smartphone usage in Indonesia is very high, reaching 270 million mobile phone numbers – a penetration rate of over 112% (Gusti, 2014).

Another core element of learning support services is the provision of Digital Library facilities. The Digital Library and its Virtual Reading Room are very instrumental in providing students with easy access to the learning materials. Complimentary to the digitization of learning materials, the Digital Library also supplies students with numerous collections of digital books and journals (Wahyono, 2014). The services through the Digital Library have been well utilized by the students – a total of 636, 585 visitors were recorded during 2014 alone.

An additional milestone for the ICT-supported learning system was the online final examination system. As known, final examinations measure the learning achievement of students. In an effort to provide a better examination service for the students, UT already had an established ICT-based Item Bank System to store its test items for easy retrieval. As students demanded more flexible examination times, in 2004 UT then started to develop a computer-based examination system that was integrated with the

Table 6.3 Total number of online self-test and students accessing online self-test Years 2011–2015

Year by faculty	Total mumber of online self-tests accessed by students					Total number of students accessing online self-tests				
	Economics	Social sciences	Education	Math & natural sciences	Total	Economics	Social sciences	Education	Math & natural sciences	Total
2011	82	108	316	136	642	19,689	15,456	31,334	8368	74,847
2012	102	109	324	138	673	35,108	25,876	53,616	13,811	128,411
2013	108	101	300	139	648	58,713	41,224	74,195	18,453	192,585
2014	143	117	142	345	747	72,402	32,368	72,871	15,782	193,423
2015	145	129	355	245	874	145,200	55,300	104,300	24,500	329,300

Source: Universitas Terbuka (2011b, 2012, 2013, 2014b, 2015)

Table 6.4 Number of courses and number of participants using the online examination system

Semester	Number of courses with online exam	Number of students/ examinees	Number of students-online exam courses
2011.1	509	966	1811
2011.2	491	1063	2123
2012.1	595	2361	4120
2012.2	592	3248	5641
2013.1	634	3848	6474
2013.2	634	4111	6180
2014.1	662	5152	7526
2014.2	674	5735	8224
2015.1	713	7342	11,444
2015.2	717	7772	12,169

Source: Universitas Terbuka (2011b, 2012, 2013, 2014b, 2015)

existing Item Bank System (Sapriati & Pardede, 2014). The implementation of computer-based examinations started in 2007, and after going through the stages of piloting and revising, in 2008 the system was converted into a fully online examination system using the university's VPN. Subsequently, students were able to take online examinations in one of the university's Regional Offices starting in 2009. The online examinations cannot be taken outside the Regional Offices for security reasons. Table 6.4 shows the number of online examination courses and the number of examinees taking the online examinations from 2011 to 2015.

As seen in Table 6.4, the number of students who take the online examination is still low compared to the overall number of students. However, it can be seen that the number of students participating in the exam online has increased from 2011 to 2015. Starting in 2015, to provide more flexibility for students to take the online exam, the exam window has been extended to 4 months, from 2 months. Furthermore, as the current online examination system is designed only for multiple-choice, and as exams must still be taken at the regional offices supervised by the proctor, UT is now developing an online examination system for essay questions. This new feature hopefully will be implemented in the near future. Similarly, UT is currently conducting research on online proctoring, so that in the future, the online examination can also be taken outside the Regional Offices.

With regard to the goal of offering fully online post-graduate programs, UT started to offer fully online master's programs in 2013, namely Master of Management, Master of Public Administration, and Master of Fisheries Management. It is expected that at the end of 2017, the fully online master's programs will also be integrated with tablet-based learning materials.

Regarding the goal of providing online community services through the offering of massive open online courses or MOOCs, UT launched its first four MOOCs in 2014 through UT's website. To increase the visibility of its MOOCs, UT decided to collaborate with an Indonesian MOOC platform called Indonesia X to also offer its MOOCs through their platform. As a result, the registration for UT's MOOCs increased from a range of 80–250 participants to over 1000 participants per course. This MOOCs provision is envisioned to deliver UT's promise for community services, providing free courses to the masses.

In accordance with the target of having all online learning supports accessible through a variety of communication devices, including mobile devices, UT continuously explores the different devices for extension of services. By the end of 2013, the main online service, that is the online tutorial, was already accessible through mobile devices. Based on a survey conducted by Padmo, Idrus, and Ardiasih (2015), it was revealed that 94% of the surveyed students often accessed online tutorials through their mobile devices (laptops, tablets, and smartphones). This data reinforces the belief that to enhance the optimal utilization by students, online learning services should be made available using various devices including mobile devices.

5 Sustainability of ICT–Supported Innovation Policies

Sustainability is one of the key principles for implementing ICT innovations at UT, as explicitly stated in the strategic and operational plans. This is to ensure that policies and programs of innovation in the use of ICT in both the administrative and academic aspects will take place continuously. Therefore, the stages of conceptualizing, developing, implementing, and evaluating are emphasized for all innovations at UT. The preparation of human resources is included in all these stages through various training programs for both the developers and users. For this, UT allocates sufficient funds for the whole process including for procurement and maintenance of ICT hardware and various applications, for various training programs, and procurement of any outsourced developments.

5.1 Human Resource Training and Development

As mentioned above one of the essential aspects to ensure the sustainability of ICT implementation in UT is the human resources training and development. These activities are carried out for all UT's human resources associated with the use of various ICT applications, as outlined in the operational plans both for administration and learning system.

A number of training activities associated with the use of ICT are carried out consistently and continuously, and include, among others training for: online tutoring, web-based infrastructure, Linux, UX Design, search engine optimization, service-oriented architecture, web design, Drupal programming, Data Quality Control, student records system, video conferencing system, optical fiber, and creating interactive multimedia. These training programs were carried out to ensure the readiness of UT's staff or external parties, who help UT to provide services or develop various computer applications. These types of training programs are and will be conducted regularly to ensure the sustainability of ICT-supported programs and services.

5.2 *Budget Allocation for ICT Development and Implementation*

Funding is an essential consideration in ensuring sustainability. UT has been consistently allocating sufficient funds for the purpose of ICT-related innovations and solutions, starting from the development of the grand design of ICT UT, the development of the applications, procurement of hardware and services related to the implementation of ICT, the cost of operations, such as annual payments of Internet network and communications, as well as funding for training and preparation of human resources in order to take advantage of related applications. Table 6.5 shows the budget allocation for ICT development and implementation for the 3 years 2013 to 2015.

As shown in Table 6.5, the budget allocation for the development and implementation of ICT-related programs is considerably adequate, with an average total allocation around IDR32.7 billion (around USD2.4 million) per year. As the table shows, a substantial amount is allocated for the cost of Internet connection, procurement of hardware and software, and maintenance of all ICT-related equipment and networking. In addition, the budget allocation for training is also adequate ensuring the sustainability of innovation in the utilization of ICT.

Table 6.5 Budget allocation for ICT development and implementation in Indonesian Rupiah (IDR)[a]

No	Activities	2013	2014	2015
1	Development of grand design ICT UT			3,210,400,000
2	Consultant for computer system design, database, computer programming, computer network	540,000,000	486,000,000	–
3	Procurement of ICT equipment	15,000,000,000	12,500,000,000	12,500,000,000
4	Computer maintenance	4,028,971,000	5,996,950,000	6,177,950,000
5	Establishment of Call Center	–	4000,000,000	–
6	Development of TIER 3 UT's Data Center	2,750,000,000	–	–
7	Internet connection	9,572,640,000	10,000,000,000	10,000,000,000
8	Training on ICT	536,179,000	535,059,000	311,642,000
	Total	32,427,790,000	33,518,009,000	32,199,992,000

[a]1 US$ = 13.000 IDR

6 Closing Remarks

The challenges of providing education to a large population (millions) scattered across nearly 17,000 islands are innumerable. ICT are at the core of the open education movement, as they are thought to break down various barriers or critical impediments to the access to quality education. These impediments are typically socioeconomic and cultural in nature and disproportionately affect the poorest, such as those who are geographically isolated. While open universities have the potential in developing countries to provide opportunities for accessing education, issues of quality, equity, and sustainability remain. Policies must be designed and enacted that address these issues in systematic ways.

In a distance learning institution with over 400,000 students and with a wide scope of services throughout Indonesia and abroad, ICT becomes a necessity. At Universitas Terbuka, innovation in relation to the use of ICT is based on the ICT grand design and the university's Strategic Plan for 2010–2021 (Universitas Terbuka, 2011a, 2011b, 2014a). Further specification of each innovation is elaborated in the university's various operational plans that provide stages of annual targets of achievement. It is clear that the university's policies for ICT utilization are to improve the quality of learning support services to students. The stages of development and implementation of all activities are well formulated in the operational plans, and the commitment to execute the plans are closely monitored and measured using the stated indicators. All this is to ensure that the policies will produce sustainable innovations in terms of ICT utilizations.

References

Belawati, T. (2014). Open education, open education resources, and massive open online courses. *International Journal of Continuing Education and Lifelong Learning*, *7*(1), 1–15.

Belawati, T., Hardhono, A. P., Sinar, E. T., & Patria, L. (2009). Development in the utilization of information and communication technology at Universitas Terbuka. In A. Zuhairi & E. Nugraheni (Eds.), *Universitas Terbuka: A journey towards a leading open and distance education institution* (pp. 1984–2008). Jakarta, Indonesia: Penerbit Universitas Terbuka.

Central Intelligence Agency (CIA). (2016). *The world factbook*. Retrieved from https://www.cia.gov/library/publications/the-world-factbook/

Damayanti, N. S. (2014). Administrasi keuangan berbasis teknologi informasi. In D. T. Belawati, N. S. Damayanti, & K. A. Puspitasari (Eds.), *Universitas Terbuka di era informasi*. South Tangerang, Indonesia: Universitas Terbuka.

Daryono, D., & Belawati, T. (2013). Prospects and challenges for introducing open educational resources in Indonesia. In G. Dhanarajan & D. Porter (Eds.), *Open educational resources: An Asian perspective* (pp. 73–84). Vancouver, BC: Commonwealth of Learning and OER Asia.

Dhanarajan, G., & Abeywardena, S. (2013). Higher education and open educational resources in Asia: An overview. In G. Dhanarajan & D. Porter (Eds.), *Open educational resources: An Asian perspective* (pp. 3–18). Vancouver, BC: Commonwealth of Learning and OER Asia.

Gusti. (2014, March 10). *Menkominfo: 270 Juta Pengguna Ponsel di Indonesia*. Retrieved from http://ugm.ac.id/id/berita/8776-menkominfo%3A.270.juta.pengguna.ponsel.di.indonesia

Hinostroza, J. E., & Labbé, C. (2011). *Policies and practices of educational computing in Latin America and the Caribbean* [Políticas y prácticas de informática educativa en América Latina y el Caribe]. Santiago, Chile: CEPAL.

Indahesia.com. (2017). *Map of Indonesia* [Online image]. Available from http://indahnesia.com/Indonesia/Indonesia.php

Internet Worldstat. (2016). *Internet usage in Asia* [Data file]. Available from http://www.internet-worldstats.com/stat3.htm#asia

Kozma, R. B., & Vota, W. S. (2014). ICT in developing countries: Policies, implementation, and impact. In *Handbook of research on educational communications and technology* (pp. 885–894). New York, NY: Springer.

Learmonth, R. (1996). IT in teaching experimental science: The scientific perspective. In D. Chambers (Ed.), *Proceedings of Dry Labs Workshop* (pp. 28–31). Sidney, Australia: University of Sydney. UniServe•Science.

McKinsey Global Institute. (2016).. *Digital Europe: Realizing the continent's potentially*. Available from http://www.mckinsey.com/business-functions/digital-mckinsey/our-insights/digital-europe-realizing-the-continents-potential

Milwan. (2014). Registrasi dan pembayaran online. In T. Belawati, N. S. Damayanti, & K. A. Puspitasari (Eds.), *Universitas Terbuka di era informasi*. South Tangerang, Indonesia: Universitas Terbuka.

Moodle. (2016). *Moodle Learning Management System* [Software]. Available from https://moodle.org/about/

Padmo, D., Idrus, O., & Ardiasih, L. (2015). *Laporan penelitian: Pemanfaatan perangkat mobile dalam meningkatkan akses terhadap pembelajaran online bagi mahasiswa universitas terbuka (The use of mobile learning in enhancing the access to online learning for Universitas Terbuka's students)*. Jakarta, Indonesia: Universitas Terbuka.

Patria, L. (2014). Komunikasi dan koordinasi. In D. T. Belawati, N. S. Damayanti, & K. A. Puspitasari (Eds.), *Universitas Terbuka di era informasi*. South Tangerang, Indonesia: Universitas Terbuka.

Republic of Indonesia. (2003). *Act of the Republic of Indonesia on National Education System* [Act No. 20, Ministry of Education and Culture]. Jakarta, Indonesia: Author.

Republic of Indonesia. (2016). *Constitution of the Republic of Indonesia 1945 (with amendments)*. Jakarta, Indonesia: Parliament of the Republic of Indonesia.

Sapriati, A., & Pardede, T. (2014). Ujian online. In D. T. Belawati, N. S. Damayanti, & K. A. Puspitasari (Eds.), *Universitas Terbuka di era informasi* (pp. 277–248). South Tangerang, Indonesia: Universitas Terbuka.

Sinar, T. E. A. (2014). Perkembangan pemanfaatan teknologi informasi dan komunikasi UT. In D. T. Belawati, N. S. Damayanti, & K. A. Puspitasari (Eds.), *Universitas Terbuka di era informasi* (pp. 52–64). South Tangerang, Indonesia: Universitas Terbuka.

United Nations Development Programme (UNDP). (2015). *Human development report*. New York, NY: Author.

Universitas Terbuka. (2011a). *Open University Strategic Plan 2010-2021: Operational Plan 2010-2013 (Rencana Strategis 2010-2021: Rencana Operasional 2010-2013, Universitas Terbuka)*. Jakarta, Indonesia: Universitas Terbuka.

Universitas Terbuka. (2011b). Annual report of the rector of Universitas Terbuka 2011. Unpublished document.

Universitas Terbuka. (2012). Annual report of the rector of Universitas Terbuka 2012. Unpublished document.

Universitas Terbuka. (2013). Annual report of the rector of Universitas Terbuka 2013. Unpublished document.

Universitas Terbuka. (2014a). *Open University Strategic Plan 2014-2021: Operational Plan 2014-2017 (Rencana Strategis 2014-2021: Rencana Operasional 2014-2017, Universitas Terbuka)*. Jakarta, Indonesia: Universitas Terbuka.

Universitas Terbuka. (2014b). Annual report of the rector of Universitas Terbuka 2014. Unpublished document.

Universitas Terbuka. (2015). Annual report of the rector of Universitas Terbuka 2015. Unpublished document.

Wahyono, E. (2014). Layanan perpustakaan digital UT dalam perspektif sejarah. In D. T. Belawati, N. S. Damayanti, & K. A. Puspitasari (Eds.), *Universitas Terbuka di era informasi* (pp. 280–302). South Tangerang, Indonesia: Universitas Terbuka.

World Bank. (2016). *Data indicators*. Retrieved from http://data.worldbank.org/indicator

Zuhairi, A., Adnan, I., & Thaib, D. (2007). Provision of student learning support services in a large-scale distance education system at Universitas Terbuka, Indonesia. *Turkish Online Journal of Distance Education-TOJDE, 8*(4), 44–64.

Author Information

Dewi A. Padmo
Universitas Terbuka
Institutional address: Jl. Cabe Raya, Pondok Cabe, Pamulang, Tanggerang Selatan 15418. Banten Indonesia

Complete mailing address: Jl. Cabe Raya, Pondok Cabe, Pamulang, Tanggerang Selatan 15418. Banten Indonesia
Telephone number: +62 21 7490941
Fax number (optional): +62 21 7403595
Email address: dewi@ecampus.ut.ac.id ; dewi@ut.ac.id
Website: www.ut.ac.id

Dewi Padmo is currently Vice Rector for Finance and General Affairs of the Universitas Terbuka (UT). She got her Masters of Arts in Educational Technology at Concordia University in Montreal, Canada. Upon her return to UT in Indonesia in 1992, she was appointed as Head of Multi Media Production Center-UT. Following from her work at the Center for Multi-Media Production, Padmo served as Head of Institute for Learning Material Development, Examination, and Information System in 2005 to 2008. Padmo got her Doctor of Philosophy in Instructional System at Florida State University in 2012. Upon returning to Indonesia, in 2012–2014 she was assigned as the Head of Institute for Research and Community Services of UT. Her research interests are mostly on teamwork and the use of technology in Open Distance Learning ODL. In the last couple of years, she has been involved in a number of research projects on mobile learning.

Tian Belawati
Universitas Terbuka
Jl. Cabe Raya, Pondok Cabe, Pamulang, Tanggerang Selatan 15418. Banten Indonesia

Complete mailing address: Jl. Cabe Raya, Pondok Cabe, Pamulang, Tanggerang Selatan 15418. Banten Indonesia
Telephone number: +62 21 7490941
Fax number (optional): +62 21 7403595
Email address: tbelawati@ecampus.ut.ac.id
Website: www.ut.ac.id

Tian Belawati is the former Rector of Universitas Terbuka and a former President of the International Council for Open and Distance Education (ICDE). She holds a Ph.D. in educational Studies from the University of British Columbia. She is also a Doctor Honoris Causa in Letters at Wawasan Open University, Malaysia. From 2008 to 2010, she was successively the Secretary General and the President of the Asian Association of Open Universities (AAOU). She won the distinguish Individual Promoter Award in 2014 from the African Council for Distance Education and is an Honorary Professor at Tianjin Radio and TV University, China.

Part III
Multistakeholder Partnerships for Local Curriculum Needs and Interests

Chapter 7
Playful Partnerships for Game-Based Learning in International Contexts

Jeremiah H. Kalir, Michael Fahy, Jeff Kupperman, Farrah M. Schiff, and Jeff Stanzler

Abstract This chapter examines the playful design and enactment of games for learning and details generative and provocative attributes of playful multistakeholder partnerships for information and communications technology-supported game-based learning in international contexts. Drawing upon the experiences of two educational technology research and design groups, the chapter first identifies four design principles that have guided various global – and playful – game-based learning partnerships in Jamaica, Oman, South Africa, and Switzerland. Such international partnerships were designed as opportunities to co-construct game-based learning by articulating the permitted, establishing collaborative presence, attending carefully to trust, and fostering third space. The second half of this chapter features a case study of ICT-supported game-based learning in Oman and describes different ways in which a playful multistakeholder partnership can be enacted in a cross-cultural setting. The subject of this case study is Place Out Of Time (POOT), a trans-historical simulation of a trial in which students play guests who come from

The original version of this chapter was revised. An erratum to this chapter can be found at DOI 10.1007/978-3-319-67657-9_9

J.H. Kalir (✉)
School of Education and Human Development, University of Colorado Denver,
Campus Box 106, 1380 Lawrence Street, Suite 645, Denver, CO 80204, USA
e-mail: remi.kalir@ucdenver.edu; http://www.remikalir.com/

M. Fahy • J. Stanzler
School of Education, University of Michigan,
610 E. University Ave., Ann Arbor, MI 48109, USA
e-mail: michfahy@umich.edu; stanz@umich.edu; http://ics.soe.umich.edu/

J. Kupperman
University of Michigan-Flint, 303 E. Kearsley St., Flint, MI 48502, USA
e-mail: jkupp@umich.edu; http://www.jkupp.com/

F.M. Schiff
Interactive Communications & Simulations, School of Education, University of Michigan,
Ann Arbor, MI, USA
e-mail: fmschiff@gmail.com

I.A. Lubin (ed.), *ICT-Supported Innovations in Small Countries and Developing Regions*, Educational Communications and Technology: Issues and Innovations,
DOI 10.1007/978-3-319-67657-9_7

a range of places and times throughout history to discuss some of the great issues of humankind and to bring the wisdom of history to a modern-day problem. Utilizing narrative methods, four vignettes from playing POOT in Oman are presented that convey the complex and sometimes contradictory characteristics of playful partnerships for game-based learning in a developing region. The chapter concludes by arguing for a more critical playfulness in game-based learning that can support all partners and players in confronting biases, celebrating difference, and creatively addressing local and global needs.

Keywords Design narrative • Design principles • Game-based learning • Oman • Partnerships • Play

1 Introduction

The spread of information and communications technologies (ICT) to improve educational opportunity in small countries and developing regions must be accompanied by an invigorated focus that simultaneously introduces and supports innovative pedagogies and social practices. Such an approach recognizes the fluid *sociotechnical landscape* of technology use in developing nations (de Laet & Mol, 2000) and also honors complex relations among social, material, cultural, and historic influences that shape the use of learning technologies for more equitable educational futures (Facer, 2011; Johri, 2011). We contend that a sociotechnical stance is necessary given international educational development efforts, which tend to foreground the primacy of tools over associated practices. Despite well-intentioned concern for a global digital divide and the inequitable distribution of educational resources, focusing foremost on access to particular tools may inadvertently ignore the promise of local social and cultural practice (and may, in fact, exacerbate entrenched inequality). Because such a dynamic may be pronounced in small and developing regions, we argue that a coupling of technological and sociocultural practices can create conditions that are more favorable for innovative pedagogical approaches conducive to twenty-first century learning. In this respect, our chapter intentionally balances attention to a given technology (digital games) and related social practices (those of game play). Moreover, we advocate that it is precisely because of complex relations among culture, power, and authority that sociotechnical practices associated with play are a promising approach to educational development work in small countries and developing regions.

Play is strongly associated with both formal and informal learning, as evidenced by connections among play and language development, socialization, problem solving, cultural identity, and creative expression (Edwards, 2000; Kangas, 2010; Lemke, Lecusay, Cole, & Michalchik, 2015; Paley, 1992; Smith, 2010). Play is also closely coupled to ICT given advances in digital games and game-based learning (Gee, 2007; Salen, 2008; Salen Tekinbas, Gresalfi, Peppler, & Santo, 2014; Squire, 2011; Steinkuehler, Squire, & Barab, 2012). There are distinct affordances and limitations when coupling together play, digital media, and game-based learning, particularly when including diverse stakeholders in the co-design of ICT and related learning opportunities (e.g. Salen Tekinbas, 2014; Squire, 2010; Williams-Pierce, 2016). Digital game play has been found to support a range of learning practices and outcomes as afforded by ICT, from inquiry and motivation to assessment and achievement (Barab, Gresalfi, & Ingram-Goble, 2010; Fishman, Riconscente, Snider, Tsai, & Plass, 2014; Wouters & Oostendorp, 2013; Young et al., 2012).

Given the scope of this volume and focus of our research, we are concerned with the adoption of digital game-based learning in international contexts and developing regions (e.g. Hwang & Wu, 2012; Schmitz, Felicia, & Bignami, 2015). More specifically, the proliferation of such media practice has both expressly and inadvertently positioned digital games as an example of ICT for development (ICTD). For example, Brewer et al. (2005) suggest that games may be a culturally appropriate means of multimedia storytelling for children in rural and low-income schooling contexts. Within content-specific domains like literacy and mathematics, researchers in India have successfully supported youth English-language learning in out-of-school contexts (Kam, Kumar, Jain, Mathur, & Canny, 2009) and have also reinforced mathematics skills through remedial education (Banerjee, Cole, Duflo, & Linden, 2007). In light of such efforts, it may not be surprising that Patra, Pal, and Nedevschi's (2009) ICTD state of the union reported that over a quarter of surveyed ICTD experts ranked games as an area of future importance in education.

Yet in spite of a small and notable research strand that has explored the role of (digital) games in social and economic development activities, Fisher (2016a) cautions that it "remains unclear whether games present a radically new approach to development, or simply a new technological platform for replicating traditional projects" (p. 33). As for games presenting a novel form of social and economic ICTD, Kolko and Putnam (2009) suggest games and gaming culture may be harnessed as one aspect of a broad *ICT ecology* that "diffuses throughout and affects nascent computer users in developing regions" (p. 47), such as in Central Asia. In this regard, games are perceived as a favorable *first touch* and *entry point* for the diffusion of ICT given low barriers to access, high motivation, and everyday relevance. Alternatively, research by Kam et al. (2008) represents the latter perspective, whereby carefully iterated digital games are deployed as a means to support traditional literacy and language learning objectives among rural children in India (see also Pal, Pawar, Brewer, & Toyama, 2006). Evidence suggests that digital games are often used simply to replicate conventional ICTD, bolstering Fisher's (2016b) complementary criticism of gamification efforts as yet another iteration of technological dominance favoring capitalist development goals. When game design and play are co-opted for explicitly productive and commercially fruitful non-game contexts, games may be critiqued for continuing a lineage of mass communication technologies that exacerbate power disparities and advance neoliberal agendas in the international development industry (Arora, 2012).

While digital games can be researched and critiqued as a tool instrumental in either transcending or (more likely) reinforcing traditional development activities, a handful of researchers have advocated an alternative focus on the promise of play in ICTD. Chirumamilla and Pal (2013) argue that an affective, open-ended, and idiosyncratic perspective on play within ICTD can create an opportunity "through which the constraining, infantilizing rhetoric of the developmental optic may be broken" (p. 31). Advocating *ludic design*, or the importance of non-productive fun, Chirumamilla and Pal describe playfulness in three ICTD cases from rural India: *Namma Dhwani* (Our Voices), a community-based radio programming effort that subverted commercial broadcasting regulations to better address local needs and interests; *Digital Green*, a series of agricultural training videos that discarded prescriptive tutorials from outside experts in favor of everyday scenarios with "people who looked like them [farmers], wearing similar clothes, talking in dialects and with mannerisms relatively similar to their own" (p. 29); and *MILLEE* (Mobile and Immersive Learning for Literacy in Emerging Economies), a game-based approach to literacy learning that foregrounded youth's everyday context and cultural practices throughout all stages of the game design process. With similar sympathy to postcolonial criticism, Ferreira (2015) centers play not as an entertaining mechanism for greater development productivity, but rather as an approach with the deliberative and constructive potential to advance freedom in ICTD.

As a point of contrast to primarily capitalist endeavors, Fisher (2016a) has advanced a model of Critical Play due to the tendency for digital games to be used in international development contexts for expressly commercial causes. Such play arises through digital games "that address structural inequality as the root cause of underdevelopment and ask players to engage in critique and conceptualize solutions outside of the current system… to challenge local and global power structures through gameplay and educate players about systemic issues that exacerbate development issues" (p. 34). The research presented in this chapter aligns well with such critical perspectives on digital games and play in developing contexts, as we consider play a disruptive break in the normative state of affairs that can serve as a means "to shock, alarm, and challenge conventions" (Sicart, 2014, p. 15) in the design and enactment of ICT-supported game-based learning.

2 Game-Based Learning and Play in Oman

Our goal in this chapter is to detail the generative and provocative attributes of play and playful multistakeholder partnerships when facilitating game-based learning in the context of Oman and to consider the broader potential of such partnerships on educational practices in small developing regions. As notions and typologies of development have come under critical review in recent years (Escobar, 2008; Sachs, 2010), Oman provides an interesting counter-example to entrenched notions of a so-called developing nation – as well as the extent to which an ICT-supported educational partnership, such as the case described here, might flourish in such a context. Despite being one of the poorer Arab states in the Persian Gulf, Oman's GDP has grown continuously in the last 50 years alongside large-scale investment in national infrastructure and economic liberalization. Notably, some of the country's largest budgetary outlays have been in basic education. Oman's commitment to reducing oil dependence while increasing economic diversification has given greater urgency to the adoption of innovative education approaches as a means of preparing the next generation of Omani students for twenty-first century careers. In spite of such development, perception – from a Western perspective – of Oman's tendency toward autocratic reform may suggest little opportunity, whether practically or politically, for more playful approaches to educational practices. Whether in an Omani or Western context, it is fair to question how games and play complement (or contradict) more conventional education systems and prescriptive learning routines. Nonetheless, the case presented in this chapter indicates Omani students and educational organizations can be supportive of ICT-supported game-based learning, and that such playfulness aligns with broader narratives about the country's ongoing commitments to development, education, and innovation.

We write as researchers, teacher educators, and designers affiliated with two game-based learning research and design groups located at the University of Michigan: the Interactive Communications and Simulations Group (ICS; http://ics.soe.umich.edu), and the Institute for Innovation in Education (iiE; http://iie.icsmich.org). Founded in the late 1970s, ICS designs and supports an assortment of web-based learning technologies for primary, secondary, and higher education students and their teachers, with the goal of promoting creative expression, cultural inquiry, and civic engagement. Having engaged over 400 primary and secondary schools in 36 American states and 25 countries (some of which may be described as developing nations, such as Brazil, Bulgaria, Morocco, Oman, Russia, Tanzania, and South Africa), as well as dozens of higher education partners, the ICS portfolio of web-based ICT include games, simulations, and online social networks concerned with various interdisciplinary topics, such as conflict in the Middle East, cross-cultural exploration and exchange, trans-historical role-play, immersive narratives, public policy debate and development, and poetry. Research about students' ICS participation suggests these projects strengthen skills such as argumentation, perspective-taking, and literacy development, while also promoting more playful orientations toward traditional schooling (D'Angelo & Kline, 2012; Holden, Stanzler, Fahy, & Kupperman, 2015; Kupperman et al., 2011; Kupperman, Stanzler, Fahy, & Hapgood, 2007). Furthermore, educator participation in ICS projects has been shown to aid pedagogical decision-making, reflection upon students' socio-cognitive engagement, and the development of mentoring capabilities (deNoyelles & Raider-Roth, 2015; Killham, Tyler, Veneble, & Raider-Roth, 2014). The ICS project featured in this chapter – Place Out Of Time (POOT) – exemplifies many characteristics of (digital) games that can foster player's identity experimentation, pro-social collaboration, and civic engagement across diverse settings and cultural contexts (e.g. Gee, 2007; Salen, 2008; Squire, 2011).

After over three decades designing game-based ICT, ICS affiliates created a new organizational mechanism to further bolster the social, cultural, and global connections necessary to enact such innovative media, practices, and learning. In 2012, the Institute for Innovation in Education established a global network of designers, educators, and researchers so as to rethink and redesign learning environments and experiences, bridge a gap between academic and popular culture, and broker social and civic engagement among K-12, higher education, and workplace learning. The iiE built upon the success of nascent global partnerships and game-based learning strategies incorporated into an educational technology graduate program facilitated by ICS researchers and teacher educators (Holden, 2013). From its inception, a hallmark of iiE programming has been an ongoing series of Gatherings – events planned collaboratively with international partners, hosted at schools and universities across the globe, and facilitated with the goal of exchanging ideas and co-constructing innovations for more equitable and participatory learning. To date, Gatherings have occurred in Switzerland, South Africa, the Czech Republic, Canada, and annually in the United States; complementary work under the iiE umbrella has engaged stakeholders from schools, universities, government, non-profit organizations, and the private sector in Jamaica, Japan, and Oman.

This chapter draws upon our ICS and iiE experiences designing and playing games with a host of global partners and describes, in particular, distinct opportunities and challenges associated with one playful multistakeholder partnership in the developing international context of Oman. ICS and iiE aim to advance ICT-supported game-based and playful learning in developing regions as a response to the fact that partnership trust, collaboration, and innovation are historically and systematically less commonplace in such contexts (e.g. Borland, 2014). As such, we address two questions in this chapter: First, what design principles guide our co-creation of playful game-based learning partnerships? And second, how is a playful partnership enacted when these principles inform game-based learning in developing regions?

3 Design Principles for Playful Partnerships

Our work with ICS and iiE is as much about creating games for learning as it is about fostering playful educational partnerships. What do we mean by playful, particularly as educational partnerships are seldom cultivated or described as playful? Following Sicart (2014), we appreciate that being playful requires appropriating a context not expressly designed for play and that "playfulness glues together an ecology of playthings, situations, behaviors, and people, extending play toward an attitude for being in the world" (p. 25). In this section, we share four design principles that guide how we *glue together an ecology* – that is, how we create the conditions – for playful partnerships given varied ICT, social practices, and cultural diversity. Heeding Hoadley and Cox's (2009) observation that "identifying a single, reliable method for learning technology design amounts to a futile search for the Holy Grail" (p. 21), we articulate some principles for playful partnership, rather than definitive prescriptions for material objects or technical processes. These principles are neither a linear and exhaustive method, nor are they abstract propositions. To avoid such interpretation, we highlight extensively our prior work in various international and game-based learning contexts as evidence of how specific design principles were developed and have evolved. In doing so, we acknowledge that our design principles may be particularly well-suited to the unique challenges and opportunities associated with ICT-supported learning in developing regions (see Lubin, Chap. 1, this volume). We then follow this section with a case study of

ICT-supported game-based learning in Oman – a learning context, we will argue, that remains in a state of development – illustrating holistically and in detail how our design principles can be enacted, at other times contested, and can lead to expressions of play, partnership, and critical playfulness.

3.1 Articulating the Permitted

Consistent with user-centered design processes that welcome a plurality of contributions during ideation and iteration (e.g. Brown, 2008), we approach game-based learning partnerships by articulating what is permitted. Among our influences as designers are (digital) games in which players' decisions are constrained by a basic premise – certain actions are required, some forbidden, and then others are permitted (e.g. Allen, Jackson, Ross, & White, 1978). We adapt this particular game mechanic as a design principle. It is the permitted – that which represents possibility and the manifold potential of play as free movement within structure (Salen & Zimmerman, 2004) – that serves as a root metaphor when we cultivate partnerships for game-based learning, particularly with partners from varied cultures, contexts, and geographies.

We neither demand, for example, that a partner purchase proprietary software as a requirement, nor do we prohibit focusing upon a partner's local curricular needs; that is, we do not emphasize our agenda while rendering theirs forbidden. We contend this is a particularly salient approach in developing regions, where resources may be scarce and where there is often "uncritical international transfer of educational policy and practice from one context to another" (Crossley & Watson, 2009, p. 636). Rather, we articulate what is permitted, such as how (digital) game play might honor players' cultures and interests (e.g. Peppler & Kafai, 2010), encourage expressions of empathy (e.g. Bachen, Hernández-Ramos, & Raphael, 2012), or embrace the ambiguity inherent to play (Sicart, 2014). When, for example, we designed DevInfo GameWorks (DIGW), an online platform that supported players' creation and exchange of games about international development data, our partnership with the Community Systems Foundation was more launchpad than contract. That relationship brokered subsequent game-based learning collaborations in developing regions – such as a partnership with UNICEF and the Jamaican Ministry of Education, and also with primary schools in South Africa – all of which were intended to meet local education needs:

> When DIGW was first proposed, the immediate goal was to increase awareness of global issues in general and the Millennium Development Goals in particular. An early pilot with urban sixth-grade students established that DIGW can be an effective tool for this aim. Since then, however, the system has been used to achieve a wide variety of different content goals: for example, a primary school teacher in South Africa was attracted by the opportunity for learners to practice English language expression in an authentic environment; a college professor used it to help her students learn literacy education concepts; and a high school journalism teacher saw it as a way for her students to do grass-roots journalism. (Kupperman, Robertson, & Baglin, 2010, p. 297)

By articulating the permitted, we privilege a more expansive approach to the co-construction of novel practices and ways of knowing, rather than market a rigid and often arbitrary codification of what counts as game-based learning and how people should (and should not) play in diverse cultural contexts.

3.2 *Establishing Collaborative Presence*

As a complement to articulating the permitted, we contend that the successful design of playful partnerships also requires establishing collaborative presence. Presence is a complex and contested notion, with social and spatial dimensions (e.g. Aragon, 2010), pedagogical implications (e.g. Anderson & Dron, 2010), and varied methodological interpretations (e.g. Slater, 1999). Our approach to presence is, first and foremost, embodied. While many of our games utilize dedicated ICT, such as websites and online social networks, our partnerships are enacted through physical and material dimensions. Prior to play, we traverse oceans, exchange gifts, and participate in everyday rituals that broker familiarity and trust in a participatory community (Lave & Wenger, 1991). Such activities demonstrate why presence is particularly important when enacting collaborative ICT-supported learning in developing regions (e.g. David, Sabiescu, & Cantoni, 2013). Though we appreciate how ICT can afford greater access and availability in an increasingly connected world, we believe that playing together means being beside one another in-game and in-room (Stevens, Satwicz, & McCarthy, 2008).

Beginning in 2007, we started experimenting with collaborative presence and agile partnership models through our facilitation of an educational technology graduate program located in Geneva, Switzerland. Our approach to the design of graduate education echoed our design of games for learning. Whether with game play or cross-cultural partnership, synchronous presence in shared space and time can deepen qualities of collaboration and make learning relevant to local needs and interests (e.g. Barron, Gomez, Pinkard, & Martin, 2014; Gutiérrez & Penuel, 2014). Many of our graduate students were classroom teachers who, collectively, called nine countries home; accordingly, their collaborative presence in the creation of learning technologies included:

> [working] with the world's leading fair labor watchdog to design and implement an online economic justice simulation for high school students. Websites and curricular resources about human rights in China, community gardening in Michigan, environmental conservation in Mexico, and global citizenship in the Congo were presented to – and then critiqued by – an official from the UN's International Labor Organization… [Another] team built solar photovoltaic systems, developed a curriculum to address barriers to rural electrification, and handed the hardware and digital resources to a South African mobile learning organization. (Holden, 2013, p. 5)

Collaborative presence, in our experience, often includes listening attentively, identifying and mitigating bias, and honoring in-the-moment improvisations characteristic of game-based learning (e.g. Holden et al., 2014). Establishing collaborative presence is not the achieved win-state of a single game. Rather, presence in a playful partnership invariably fluctuates, with collaboration demanding ongoing negotiation among the values, practices, and objectives of all partners – a necessary, yet challenging, premise when partnering in developing contexts.

3.3 Attending to Trust

Like collaborative presence, trust in educational partnerships is neither a given nor an outcome established via formulaic procedure. Rather, attending to trust is a process of attunement whereby familiarity grows among stakeholders' cultures, norms, and expectations (e.g. López Turley & Stevens, 2015). In our work, trust is a social practice that recurrently attends to boundaries and boundary crossing. A boundary may demarcate disciplinary expertise; or suggest sameness, highlight sociocultural difference, or define what "counts as… central participation" (Akkerman & Bakker, 2011, p. 132). Boundary crossing, therefore, can refer to practices and dynamics that (re)design and (re)conceptualize learning across intercultural contexts (e.g. Resta & Laferrière, 2015), as well as among multiple perspectives and partners, demonstrating "ways in which sociocultural differences and resulting discontinuities in action and interaction can come to function as resources for learning and development of intersecting identities and practices" (Kumpulainen & Sefton-Green, 2014, p. 13). Trust in partnerships entails recognizing and respecting boundaries, from the ritual of cultural tradition to insight from personal experience to the constraints of a game as designed learning experience.

Yet attending to trust in a playful partnership also honors – and actively encourages – boundary crossing. Recall that play may be understood as both free movement within structure and also a transgression of norm or expectation. When, for instance, learners in developing nations like Jamaica and South Africa designed DIGW games, our game-based learning partnerships required trust when honoring spatial, technical, and conceptual boundary crossing (Kupperman et al., 2010). Together, children, adults, and researchers attended to the joint work of game design as movement among people and places, data, and ideas about community and identity:

> In contrast to the schools we worked with in South Africa, where the learners ventured out into the community to gather material for their games, the learners in Kingston used information they could gather from the computer lab, much of it using DevInfo software to access data about Jamaica. We expected this to be challenging, and it was, especially considering that DevInfo is a sophisticated tool used mostly (until now) by governments and UN agencies, and the learners are just completing grade six… The questions the kids came up with were important and real: Is crime in Jamaica increasing or decreasing? How much poverty is there here compared to other countries? Is there a difference between how girls and boys are doing in school? The conclusions they came to were not always what they expected at first, and with some guidance, they were able to form their findings into a game that is quite accurate, and yet entirely their own. (Kupperman, 2010, para. 3)

Whether in data analysis or game design, boundary crossing promotes different perspectives on shared activity and contributes to the co-construction of novel practices. Because play is transgressive and also transformative, we attend to trust in our game-based learning partnerships by creating the conditions for continuous joint work at and across boundaries.

3.4 Fostering Third Space

The final principle that guides our approach to playful game-based learning partnerships is fostering third space. With a rich conceptual lineage (e.g. Bhabha, 1994; Soja, 1996) and transformative enactment in educational research (e.g. Gutiérrez, 2008; Moje et al., 2004), varied definitions of third space broadly refer to the emergence of new perspectives, narratives, and practices – and for both individuals and groups – that result from hybrid discourses and spaces. Our commitment to fostering third space reflects the consequences of boundary crossing: "Given a certain problem space, practices that are able to cross their boundaries engage in a creative process in which something hybrid – that is, a new cultural form – emerges" (Akkerman & Bakker, 2011, p. 148). Because ICS games (like DIGW) and iiE efforts (like playful partnerships) span both developing and developed international contexts, we contend that initiatives like ICT-supported game-based learning are well-suited to foster hybrid and emergent learning:

> This may be the most important reason for games to be part of school. Games provide students the opportunity to break with common sense and yet be taken seriously in the context of the game, even though everyone knows that outside the game what they are doing might be absurd. Games provide us an opportunity to take our students' explorations seriously, without being judgmental or condescending. They provide safe bubbles where students (and teachers) can take risks and try out different ways of thinking, inviting others to attend to their actions while preserving a distinction between their own self and the role they have taken on. These actions are worthy of our attention because they will show whether players have been able to take the leap to a different way of thinking. If our job is to facilitate, then it is not so much to help the players understand where they have arrived as to help them understand where they have been. (Kupperman et al., 2011, p. 29)

Our approach to game-based learning partnerships in both developing (i.e. Jamaica, Oman, South Africa) and developed (i.e. Japan, Switzerland) countries eschews so-called best practices in global ICT efforts (e.g. Zhang, Yang, Chang, & Chang, 2016) that may too readily average Western perspectives with the diverse cultures of stakeholders in such contexts. On the contrary, we are cognizant that fostering third space is suitable – if not increasingly necessary – so as to advance innovative ICT partnerships in developing contexts that are divorced from a history of distrust and the interventionist diffusions of top-down reform (David, Sabiescu, & Cantoni, 2013). Our shared partnering seeks to evince something entirely distinct. What we describe as a playful partnership necessarily draws upon, and yet simultaneously seeks to transcend, the dominance and bias of any one culture or way of knowing. We maintain that playfulness through games can foster – given participatory design, trust, and collaborative partnering – a third space for nascent and transformative learning.

4 A Playful Partnership for Game-Based Learning

Having identified design principles for playful game-based learning partnerships, we now present a case study that describes different ways in which a playful partnership may be enacted in a cross-cultural and developing nation context. This case features the ICS game Place Out Of Time (POOT), a trans-historical simulation of a trial in which students play guests who come from a range of places and times throughout history to discuss some of the great issues of humankind (i.e. justice, identity), and to bring the wisdom of history to a modern-day problem (i.e. religious liberty). To do so, students role-play – online and in-person – historical and contemporary characters (e.g. Queen Isabella, Gandhi), and discuss cultural, social, and moral issues in those characters' voices. We loosely employ the motif of a trial to animate players' POOT interactions, usually designating a handful of characters as justices, who solicit support from other guests, and in the end, formally issue their opinion on a case. The process creates ample room for discussion and politicking, and in so doing, students have a contextually-rich opportunity to practice historical research, deliberation, and perspective-taking (Kupperman et al., 2007). We also want students to embrace theatrical attitudes when playing POOT, such that their social interaction – primarily via online writing – is imbued with the *voice* of their character, and the tone they imagine their character might take (Kupperman et al., 2011). Figure 7.1 is a screenshot from the POOT website that supported role-play and debate in our featured case.

4.1 Establishing POOT in Oman

In 2014, after over a dozen years orchestrating imaginary encounters among diverse historical and social perspectives, we found ourselves in a real-life meeting of cultures. Representing ICS and iiE, we were invited to design and facilitate a version of POOT for a group of secondary students in Oman. These students were participants in a summer residential program (SRP) organized by the Takatuf Scholars Programme. Takatuf is a social responsibility initiative of the Oman Oil Company, the nationalized oil enterprise of the Sultanate of Oman, and one manifestation of decades-long Omani educational reform (e.g. Al Balushi & Griffiths, 2013). Each year, a cohort of 80 Grade 11 Takatuf Scholars participate in academic and extracurricular programming. Through a competitive evaluation process, ten Scholars from among this group are awarded full scholarships to both an international preparatory school and then a four-year university in the United States, Canada, or the United Kingdom. Scholars attend the SRP for three weeks to commence their Takatuf experience. POOT's digital game-based learning and communication-intensive qualities appealed to our partners as a novel means of initiating the SRP in alignment with Oman's broader education reform efforts.

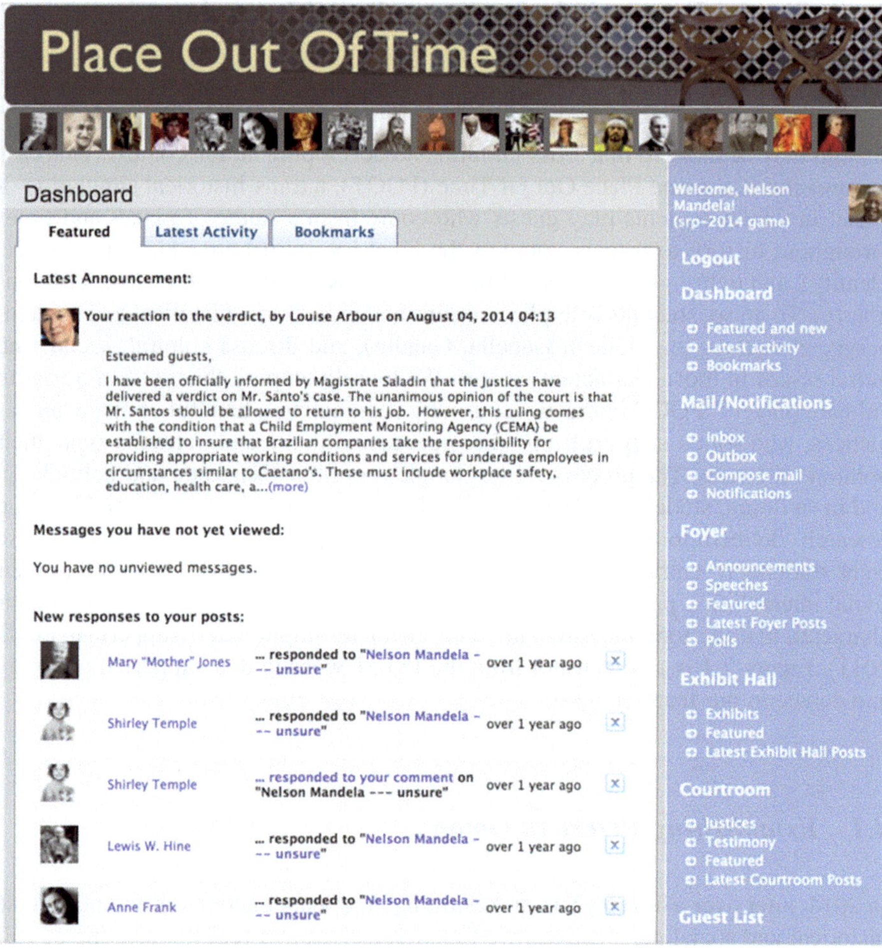

Fig. 7.1 POOT website that supported Omani game-based learning partnership

Briefly, education in modern Oman has undergone dramatic transformation across elementary, secondary, and higher education systems (Alfar, 2013; Hussain & Al Barwani, 2015) that reflect the country's overall "growth miracles" (United Nations Development Programme, 2010, p. 3). Until 1970, there were all of three schools with a modern curriculum in the entire country, with a total enrollment of 900 pupils (Al Balushi & Griffiths, 2013). Following the accession to power of Sultan Qaboos in 1970 and a series of educational reforms, that figure rapidly expanded; by 2010, over 520,000 students were enrolled in 1040 schools. All Omani children have access to a free education through the end of the secondary level. While school is not mandatory at any level, more than 90 percent of school-aged children are enrolled, with girls represented proportionately to their general population demographics. Since the 1990s, and so as to prepare "young Omanis for life and work under the new conditions created by the global economy" (p. 155), the Ministry of Education has introduced comprehensive reforms designed to prepare Omani students for a globalized knowledge economy, placing emphasis on learner-centered education, the acquisition of communication and learning skills, and the development of critical thinking (see also United Nations Development Programme, 2010).

It is within this educational reform context that our game-based learning partnership with Takatuf sought to further their goal of providing "secondary school students with the practical and intellectual tools needed to support their success not only as university students but also as lifelong learners and future leaders."[1] In addition to Takatuf, our prior work with Omanis and extensive international contacts allowed us to establish a network of stakeholders from multiple sectors to meaningfully enact POOT in Oman. Our multistakeholder partnership included: Takatuf's directors and staff; the Sultan Qaboos Cultural Center; a member of the *Diwan*, Oman's Royal Court; a representative from the International Labour Organization in Geneva, Switzerland; and a financial analyst from the Central Bank of Oman, who first became acquainted with ICS games as an undergraduate student at the University of Michigan.

A playful attitude among the many stakeholders of our partnership was necessary given a number of constraints unique to the Omani version of POOT. First, the usual online model of POOT spans eight weeks; this version would need to conclude in three days. Second, when POOT is facilitated online, we rely on classroom teacher-partners to support students as they research and become their characters; in Oman, we would need to more concretely support this process ourselves, building on work begun during the spring by Takatuf staff, and taken on individually by the Scholars in the intervening weeks. Third, the 80 participating Scholars were paired to play shared characters. Among the 40 shared characters were those reflecting cultural traditions and histories more familiar to Omani students (e.g. Ibn Rushd) as well as contemporary figures known for more controversial, if not explicitly Islamophobic, opinions (e.g. Marine le Pen). Fourth, though we drafted a scenario about migrant labor in Oman, *playing* with the volatility of regional politics was simply too risky. Accordingly, the game's central trial concerned child labor and leveraged residual excitement from the 2014 FIFA World Cup that had recently concluded in Brazil to focus on that country's child labor issues. The scenario concerned Caetano Santos, a fictional Brazilian teenager who worked in a clothing factory until fired following an investigation into illegal child labor. Caetano – proud of contributing to his family's wellbeing – sought a ruling from the POOT Court of All Time that would allow him and other minors to return to work.

Given the unique contingencies of implementing POOT in Oman, we were purposeful about the use of ICT by Scholars, Takatuf and SRP staff, and our facilitation team. For example, each Scholar had access to a laptop computer. As such, they were able to research their characters online and practice *speaking* as their characters in written peer-to-peer interactions on the POOT website. Via these online forums, Scholars could also communicate directly with Caetano, played by a member of our

[1] For additional information about the Takatuf Scholars Programme see www.takatufscholars.com

Fig. 7.2 A mock newscast that playfully introduced the Omani POOT scenario

team based in the United States. Throughout the game, we also presented a series of mock newscasts (Fig. 7.2) that provided important information about child labor in Brazil and gave Caetano a face (i.e. stock image) and voice (via internet translating services). Scholars were further immersed in the game's scenario as Takatuf staff role-played reporters and recorded video interviews of Scholars while they were in-character. Finally, we utilized video conferencing technology so that Scholars, while role-playing, could speak with an international labor expert in Geneva, Switzerland.

The following vignettes provide additional glimpses of our playful, multistakeholder partnering in the creation, facilitation, and reflection upon POOT in Oman. Influenced by narrative inquiry (Clandinin, 2006) and design narrative (Bell, Hoadley, & Linn, 2004) methods, these four vignettes deliberately play with voice (both ours as researchers, and also those of participating Takatuf Scholars), style, and structure. We contend this approach more authentically represents the complex – and sometimes contradictory – characteristics of playful partnerships for ICT-supported game-based learning in international contexts.

4.2 Vignette 1: Introductory Quote Cards

Nine o'clock on a sweltering August morning. Thirty minutes into our first day of POOT Oman and everything looks fairly traditional. Our team of four facilitators stands at the front of a classroom. We introduce ourselves and POOT to 80 Takatuf Scholars – young women seated to our right, young men to our left – listening attentively. Atop their desks is a numbered card. For our first activity, we explain, everyone will stand, move about, and find three peers with the same numbered card. Having formed a group of four, they

Fig. 7.3 Takatuf Scholars form groups during introductory Quote Cards activity

will flip over their cards, read a quote, and collectively interpret its meaning. But first, we do something quite bizarre. We four go to different corners of the room, then begin wandering about, each waving a card numbered 21. We quickly gather together. One of us reads aloud: 'I am a citizen of the world.' We pause, then ask: 'What does *that* mean?' More silence. Another one of us asks, 'Well, what does *citizen* mean?' Collectively, publicly, we analyze our quote, each offering a new interpretation. The Scholars quietly observe our modeling, their faces excited and a bit puzzled.

And then, 'Go!' The invisible line separating male and female disappears as Scholars whiz past each other, stopping only to compare cards, group, or move on (Fig. 7.3).

The groups of Scholars assemble – some unisex, some mixed – and clump together around the room, at desks, and in the hallway. They speak loudly. Dozens of excited voices exchange ideas for fifteen minutes. We facilitators then gather five Quote Card groups together and find space where each can summarize their discussion and present interpretations. Some groups do not reach a consensus, offering multiple meanings of a quote. Some have their interpretation challenged by peers. Most Scholars are eager to speak. The 20-min debrief stretches to 40. One group, for example, interprets 'It is the first responsibility of every citizen to question authority,' from Benjamin Franklin, to mean citizens should *try to understand* or *have a dialog* with authority, rather than to challenge authority. Given our Western biases, the Scholars' interpretations often challenge our perspectives and contradict our expectations.

The reflection ends, the Scholars take a break, and we debrief. We are in awe. The Scholars' eagerness to participate is noteworthy given what we ultimately learn about their day-to-day education. For a majority of the 80 Scholars, the SRP is their first mixed-gender learning opportunity. Others would also tell us how their schools follow a direct instruction model, within which they obediently listen, take notes, and do not participate in such spontaneous and open-ended discourse. Their typical schooling stands in stark contrast to the Quote Cards activity. We are also aware that the Scholars' variable English proficiency did not limit collaborative participation. Our Takatuf partners had requested a language-intensive introduction to POOT's terminology. Yet rather than study vocabulary out of context, we chose to plunge the Scholars into a playful examination of quotations, shifting the activity from definition to interpretation. The Scholars' willingness to play along is that much more impressive.

4.3 *Vignette 2: Misreading Signs*

In this version of POOT, the game's scenario features the plight of Caetano Santos, a Brazilian teenager who worked in a clothing factory until fired following an investigation into illegal child labor. Caetano – proud of contributing to his family's wellbeing – seeks reemployment and brings his case before the Court of All Time. While fictitious, our POOT scenarios and trials are always based upon real world concerns. In Brazil, anyone younger than 16 is not permitted to work, yet apprenticeship laws and inconsistent compliance practices present persistent challenges; over one million boys and girls, aged 5–14 years, work illegally in Brazil (Bureau of International Labor Affairs, 2014). Lending further credibility to the POOT scenario, we arrange a video conference so that the Takatuf Scholars can discuss – while role-playing as their characters – questions about child labor, employment, and economic security with a policy analyst at the International Labour Organization in Geneva, Switzerland. As with many well-intentioned designs, however, we confront an unanticipated challenge.

Our conversation is organized for the second morning of the SRP, less than 24 h after we begin playing POOT. All 80 Scholars sit together in our shared classroom, young women on the left, young men on the right. At the front of the room, we arrange desks and computers to facilitate our video conference and invite Scholars to sit before their peers, pose questions to our guest expert, and publicly discuss child labor (Fig. 7.4). Something, however, is amiss.

During our first day together, the Scholars confidently adapted to the playfulness of POOT and enthusiastically embraced their character role-play. While we recognize the unique mixed-gender arrangements of the SRP, the Scholars readily interacted in this atypical learning environment. Perhaps mistakenly, we begin to question our own assumptions about norms associated with education, gender, and social interaction. As the conversation linking Muscat and Geneva commences, first one male Scholar, then another, and a third sits to ask a question. We quickly realize that young men are entirely comfortable role-playing in this particular physical arrangement, and in social interaction with an expert male stranger. Alternatively,

Fig. 7.4 A Takatuf Scholar discussing child labor with ILO economic policy analyst

not a single female Scholar is willing to leave her seat and enter what is – as readily becomes clear – a highly gendered space. With this realization, we immediately adjust our facilitation by circulating a microphone to female Scholars so that they are welcomed into the conversation. Only then does our activity proceed more equitably and with greater Scholar participation.

4.4 Vignette 3: Be(com)ing Mother Jones

With thick, plastic-rimmed white glasses and Elvis-inspired hair, Zarif[2] looks like someone who can push a few boundaries. He is unhappy role-playing a woman, particularly someone with whom he is entirely unfamiliar – the American labor organizer Mary 'Mother' Jones. In typical versions of POOT, young women frequently role-play male characters, while young men rarely choose to play female characters. In Oman, strong social norms determine where and how males and females can gather together. As such, we – navigating our own roles as outsiders – do not risk assigning female characters to the male Scholars. We work carefully with our Takatuf partners to finalize an appropriate list of 40 historic and contemporary characters who can debate the game's child labor scenario. Ultimately, the characters are assigned to Scholars by Takatuf's staff. Hence, Zarif – in sideburns and his gray dishdasha robe – is challenged to become Mother Jones. And over the course of only a few days, Zarif-as-Mother Jones capably testifies during POOT's final trial. Speaking to a panel of three adults role-playing the trial's justices, Zarif confidently remarks as Mother Jones: 'In my days, I encountered many factories

[2]Zarif is a pseudonym.

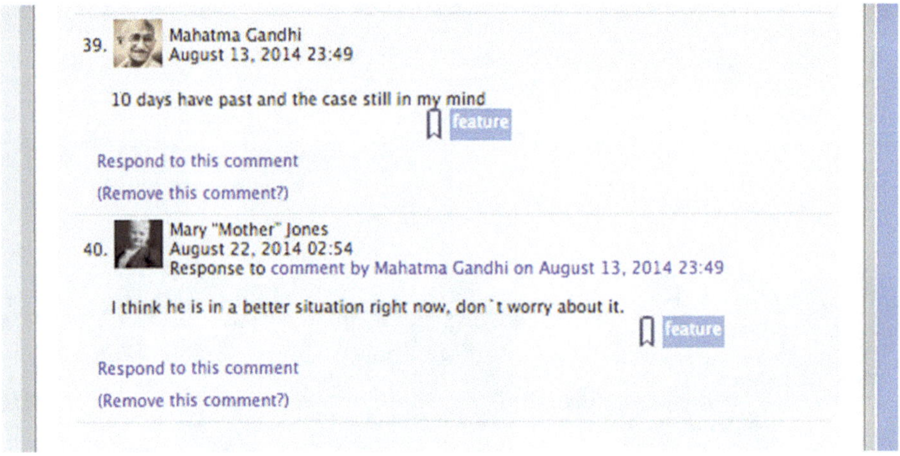

Fig. 7.5 Mother Jones' comment to Gandhi more than two weeks after POOT concluded

that hired children to work, and I succeeded in opposing most of them… With my great experience and knowledge in these factories, I'm sure that such factories are dangerous and unsuitable for children to work at.'

What allowed Zarif to convincingly role-play Mother Jones – whether with fellow Scholars face-to-face, in written online exchange, or before a panel of highly esteemed adults? And what compelled Zarif – more than two weeks after the game ended – to continue speaking as Mother Jones in response to Gandhi (Fig. 7.5)?

In a debrief, Zarif told us that despite his preconceptions, when researching Mother Jones he learned of someone who was feisty, strong, and independent-minded. Indeed, she was the kind of person he aspired to become in his life. Even more surprisingly, Zarif discovered that his own father knew of Mother Jones and was a lifelong admirer. These conditions may not entirely explain how Zarif role-played someone of a different age, culture, and gender; someone who had different life experiences and who was not afraid to challenge authority. For a young man in Oman to act similarly in everyday circumstances could have notable consequences. However, the playful constraints of POOT suspended certain cultural norms, allowing players' actions to carry new and different meanings. POOT encouraged Zarif, like other Takatuf Scholars, to play with the rules governing social interaction: 'So when I came here, at first, I thought 'How am I going to do this?' But actually, it was very easy for me. I don't know how, but I actually got into my character so much that when I'm talking from Mother Jones' perspective, I forget myself!'

4.5 Vignette 4: A Trial, and a Verdict?

I am Louisa May Alcott, from United States. I was a novelist and social worker. I do support Caetano's request. When I was a child, the poverty made it necessary to work… We need to live in a world that all people have decent, happy lives… Would

Fig. 7.6 Takatuf Scholars roleplaying Louisa May Alcott testify during trial

someone prefer to work and live a decent life, or read a book with an empty stomach in a place just like that [Scholar points to image of poverty, Fig. 7.6]? This would be Caetano's situation if he doesn't get a job.

With 80 Scholars in the same place at the same time, we amplified dramatic aspects of the game's final trial that are not possible in the larger, exclusively online version. Each pair of Scholars would jointly present their character's stance on Caetano's case, arguing in response to key questions that had guided our research and preparation. Are human rights best served if children are protected from working before they reach adulthood? Should human rights be defined as protecting the rights of young people to make their own decisions about whether or not to work? Furthermore, testimony would occur before three adults role-playing as Justices: a member of the *Diwan* (Oman's Royal Court); our former student, now an analyst at the Central Bank of Oman; and the Programme Director of Takatuf. Would this Court of All Time judge Caetano based upon economics, morality, safety, or individual liberty?

When we asked Takatuf's Programme Director to serve as a Justice she readily agreed. On the final morning of POOT, shortly before Scholars began testifying, she hastily informed us that the Justices would not render a 'yea or nay' verdict about Caetano's fictitious case. The Programme Director, herself a North American, strongly opposed any definitive outcome. Confused, we inquired about her rationale. She was not worried about the Scholars' debate or competitiveness. Rather, Scholars could plausibly regard a verdict enjoined by a member of Oman's Diwan as a formally sanctioned truth. Moreover, a Scholar's argumentative stance – even one presented during role-playing – could be construed as undermining the authority of a government official, leading to potentially inappropriate or shameful outcomes. Despite having explained role-play responsibilities, and having secured these leaders' eager cooperation as Justices, it appeared that our effort to simulate a more realistic trial with support from respected representatives of Oman's civil society was forbidden. Ambiguity and conflict notwithstanding, we played on. Scholars' testimony (Fig. 7.7) alternated between characters favoring Caetano's right to work and those arguing for his protection as a minor.

As is evident in the exchange below, the justices routinely engaged the Scholars in thoughtful and respectful deliberation.

> Nelson Mandela: I'm not saying that children should not work in factories or in a safer place. I'm saying children must not work at all. The government must provide them with support, like money, food, and shelter, so that they don't have to work until they are the proper age. Let's say the government provides Caetano's family with money. He wouldn't have to work. In fact, he would go to a school and he would educate himself so that, in ten years from now, he will have made an income for himself.
>
> Justice: We make the assumption that the government has unlimited resources. As someone that studies economics, that's not actually a fact. So when you always say, 'The government should, the government should, the government should,' we anticipate that the government has endless funds. But in some countries, the governments don't. And in those countries, the government cannot facilitate completely – or for all the children – what you're requesting. Then what do you suggest in this case?
>
> Nelson Mandela: Well, I'm saying, the country we're talking about right now is Brazil. Brazil has so much poverty. And during 2014, Brazil has spent over two billion dollars on the World Cup, right? Those two billion dollars they could have put toward fixing poverty, schools inside the country. Yes, they have limited resources, but they do have resources they could have put in education.

The Scholar's nuanced analysis, character-appropriate perspective taking, and in-the-moment improvisation were impressive. Yet throughout the testimony, we worried about how the Justice's decision could avoid conflict and – what we believed

Fig. 7.7 Takatuf Scholars roleplaying Ellen Johnson Sirleaf testify before Justices

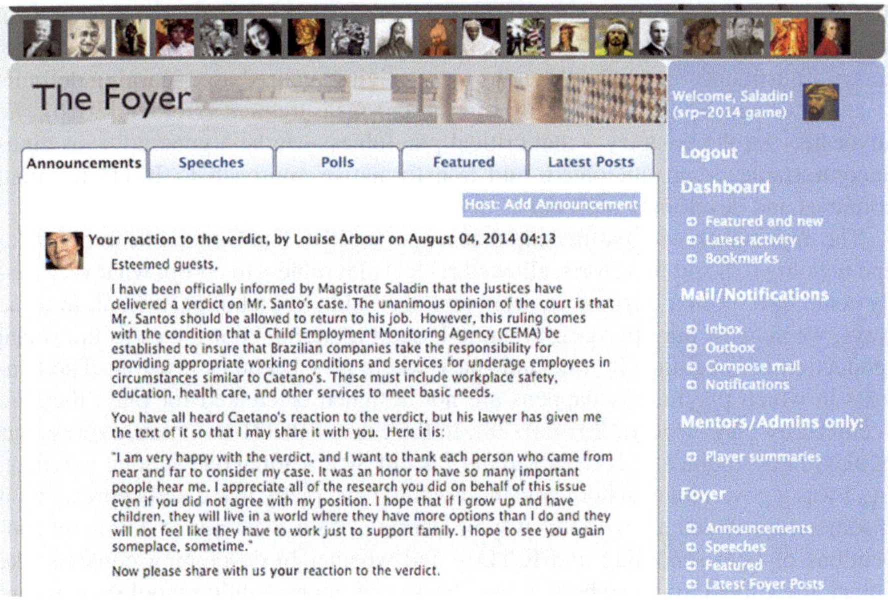

Fig. 7.8 The trial verdict posted online

would become – a dramatic anticlimax. Our uncertainty only deepened as the Justices retreated to a separate classroom for their own deliberation. And then, quite unexpectedly, they returned with a verdict. A unanimous decision determined that Caetano would return to his job at the clothing factory. The Justices presented the verdict, accompanied by the Scholars' shock and excitement; the result was also posted in detail to the game's online platform (Fig. 7.8).

We will never know what prompted the Justices to agree – unanimously – upon their verdict. Perhaps the momentum of the trial dictated this discrete outcome. Perhaps Scholars' role-play and passionate testimony rendered the verdict ancillary to their more impressive demonstration of competence and creativity. Or maybe the in-character interactions between Scholars and Justices were the truly significant occurrence. There was, we believe, an appropriate and playful ambiguity to the game's conclusion.

5 Toward Critical Playfulness in Game-Based Learning

We have described how international partnerships generally and, more specifically, multistakeholder collaboration in developing regions can be an opportunity for co-constructing activities of critical playfulness (i.e. Ferreira, 2015; Fisher, 2016a) based upon principles of articulating what is permitted, establishing presence, attending carefully to trust, and fostering a third space in which game-based and playful activity occurs. This playful approach to educational partnerships means that learning outcomes may not be completely known in advance, and this, in turn, makes it possible to take thoughtful risks, critique decision-making, and make necessary adjustments on the fly. Whereas some (digital) game-based learning delimits stakeholder agency for the sake of control and measurement (Holden, 2016), we advocate – on the contrary – that critical playfulness can be a generative, at times uncomfortable or unanticipated, and transformative approach to ICTD in small countries and developing regions.

The multistakeholder partnership that supported POOT in Oman, like our broader and ongoing ICS and iiE efforts, allowed critical playfulness to co-opt what very easily could have been a learning environment more akin to traditional school. In these ways, we suggest our approach to game-based partnering contributes to the social production of "openings in time and space where play becomes possible. The contexts in which playfulness happens are not designed or created for play: they are occupied by play" (Sicart, 2014, p. 28). In the case of POOT in Oman, expressions of playfulness momentarily occupied the Takatuf SRP, leading one Scholar to reflect, 'As I was a government school student, I would have really loved and appreciated it if someone provided me with this way of learning.' As a challenge to dominant conventions of both schooling and ICTD – and whether in developing contexts like Oman, the United States (where, it may be argued, many public school systems are similar to those in developing nations), or elsewhere – playful game-based learning partnerships can engender hybrid spaces whereby all partners, irrespective of age and status, may speak differently, playact, critique, and design new ways of learning.

Given the distinct qualities of our playful multistakeholder partnerships, it is reasonable to anticipate questions about implementation, impact, and sustainability. What happens if such partnerships are not advanced? How might individuals and organizations appreciate the impact and meaning of critical playful learning? And are such playful partnerships sustainable given more traditional approaches to schooling and learning? As designers and researchers of learning technologies – and not, for that matter, experts in comparative education or conventional international education development – we suggest such questions are best addressed by further amplifying the voices of our partners. We contend it is important to advance such partnerships because of learners who, after only a few days playing POOT, noted: 'The main thing I learnt is that we shouldn't just look at one perspective. In fact, if we take everyone's opinion we might end up with a solution to a problem that everyone is happy with. This event has also given me multiple skills like teamwork, listening, decision making and how to put myself in others' shoes.' Moreover, in the view of Scholars, such partnerships should be sustained as complementary to current educational practice: 'I think this event should be applied in school at least once a week every two months or so. It gives students so many skills that they will be able to use in the future; be it at school, university, work or even dealing with people everyday.'

Because we are designers and educators who regularly traffic along and across boundaries, we remain ever-curious – if not also skeptical – about the types of impact and meaning-making that accompany ICTD and ICT-supported game-based learning. We certainly appreciate the Takatuf Scholar who remarked, 'I learned so many things such as how to see things from another person's perspective, embrace, and accept other people's personalities.' Yet we cannot simply congratulate ourselves on a game well played. Indeed, many other game-based learning opportunities scaffold players' perspective-taking (e.g. Steinkuehler et al., 2012) or promote distinctive partnerships (e.g. Salen Tekinbas et al., 2014). Accordingly, we contend that cross-cultural and multistakeholder play – rife with excitement, miscommunication, and surprise – may serve more complex and expressly critical purposes. In this respect, we recognize how "games serve as tools to make privileges, positions, and points of view visible" (Souto-Manning, 2011, p. 999). Furthermore, we admire Flanagan's (2009) orientation to critical play as gameplay "characterized by a careful examination of social, cultural, political, or even personal themes that function as alternates to popular play spaces" (p. 6). These perspectives are both rare and necessary – particularly in ICTD contexts – given how game play so easily traverses fluid and expansive social networks. So too are these perspectives indispensable if game-based learning is to more meaningfully help players and learners from around the world confront biases, celebrate difference, and creatively address local and global needs.

We will always carry our Western positions and personalities into international game-based learning partnerships. As such, our design principles for playful partnerships and our commitment to critical playfulness must serve multiple ends for all stakeholders: establish a shared educational endeavor through collaboration and mutual respect; create the conditions for participatory and more equitable game play; and support players' critique of social, cultural, and historical perspectives. Moreover, both our design principles and game-based learning opportunities must also function as a cohesive set of strategies for mitigating the harmful effects of unquestioned power, privilege, and global inequality. Playing POOT in Oman, we learned that boundaries are not always clearly marked (including our own), that signs can be both poorly designed and misread, and that at times certain assumptions need not be discarded but rather carefully refined. Wherever gameplay takes ICS and iiE – across geographies, concepts, and cultures – we recurrently affirm that a particular type of critical and playful agility is something we seek to cultivate in learners, in partners, and in ourselves.

Acknowledgements We thank the many students, classroom teachers, and education leaders on five continents who have participated in ICS and iiE games and programming over the past 30 years. We are particularly grateful to the students and staff of the Takatuf Scholars Programme, as well as our other colleagues in Oman, without whom the playful partnership described in this chapter would not have been possible. Lastly, we thank Fred Goodman, whose work continues to be the foundation – and an inspiration – for all that we do.

References

Akkerman, S. F., & Bakker, A. (2011). Boundary crossing and boundary objects. *Review of Educational Research, 81*(2), 132–169.

Al Balushi, S., & Griffiths, D. (2013). The school education system in the Sultanate of Oman. In G. Donn & Y. Al Manthri (Eds.), *Education in the broader Middle East: Borrowing a baroque arsenal* (pp. 107–126). Oxford, UK: Symposium Books Ltd..

Alfar, E. (2013). Oman. In D. Ness & C.-L. Lin (Eds.), *International education: An encyclopedia of contemporary issues and systems* (pp. 657–659). New York, NY: Routledge.

Allen, E., Jackson, G., Ross, J., & White, S. (1978). What counts is how the game is scored: One way to increase achievement in learning mathematics. *Simulations and Games*, *9*(4), 371–392.

Anderson, T., & Dron, J. (2010). Three generations of distance education pedagogy. *The International Review of Research in Open and Distributed Learning*, *12*(3), 80–97.

Aragon, S. R. (2010). Creating social presence in online environments. In S. R. Aragon (Ed.), *New directions for adult and continuing education* (pp. 57–68). San Francisco, CA: Jossey Bass.

Arora, P. (2012). The leisure divide: Can the 'Third World' come out to play? *Information Development*, *28*(2), 93–101.

Bachen, C. M., Hernández-Ramos, P. F., & Raphael, C. (2012). Simulating REAL LIVES: Promoting global empathy and interest in learning through simulation games. *Simulation & Gaming*, *43*(4), 437–460.

Banerjee, A. V., Cole, S., Duflo, E., & Linden, L. (2007). Remedying education: Evidence from two randomized experiments in India. *The Quarterly Journal of Economics*, *122*(3), 1235–1264.

Barab, S. A., Gresalfi, M., & Ingram-Goble, A. (2010). Transformational play using games to position person, content, and context. *Educational Researcher*, *39*(7), 525–536.

Barron, B., Gomez, K., Pinkard, N., & Martin, C. K. (2014). *The Digital Youth Network: Cultivating new media citizenship in urban communities*. Cambridge, MA: The MIT Press.

Bell, P., Hoadley, C. M., & Linn, M. C. (2004). Design-based research in education. In M. C. Linn, E. A. Davis, & P. Bell (Eds.), *Internet environments for science education* (pp. 73–85). Malwah, NJ: Lawrence Erlbaum Associates.

Bhabha, H. (1994). *The location of culture*. London, UK: Routledge.

Borland, R. (2014). The PlayPump. In S. Walz & S. Deterding (Eds.), *The gameful world: Approaches, issues, applications* (pp. 323–338). Cambridge, MA: MIT Press.

Brewer, E., Demmer, M., Du, B., Ho, M., Kam, M., Nedevschi, S., … Fall, K. (2005). The case for technology in developing regions. *Computer*, *38*(6), 25–38.

Brown, T. (2008). Design thinking. *Harvard Business Review*, *86*(6), 84–92.

Bureau of International Labor Affairs. (2014). *Brazil: 2014 findings on the worst forms of child labor*. Washington, DC: United States Department of Labor.

Chirumamilla, P., & Pal, J. (2013, December). Play and power: A ludic design proposal for ICTD. In *Proceedings of the Sixth International Conference on Information and Communication Technologies and Development: Full Papers-Volume 1* (pp. 25–33). New York, NY: ACM.

Clandinin, D. J. (Ed.). (2006). *Handbook of narrative inquiry: Mapping a methodology*. Thousand Oaks, CA: Sage Publications.

Crossley, M., & Watson, K. (2009). Comparative and international education: Policy transfer, context sensitivity and professional development. *Oxford Review of Education*, *35*(5), 633–649.

D'Angelo, J., & Kline, S. L. (2012). *Argumentative discourse skill development in online educational simulations: How George Carlin can teach critical thinking*. Paper presented at the American Educational Research Association Annual Meeting, Vancouver, BC.

David, S., Sabiescu, A., & Cantoni, L. (2013, November). Co-design with communities. A reflection on the literature. In J. Steyn & A. G. Van der Vyver (Eds.), *Proceedings of the 7th International Development Informatics Association Conference* (IDIA) (pp. 152–166). Bangkok, Thailand.

de Laet, M., & Mol, A. (2000). The Zimbabwe bush pump: Mechanics of a fluid technology. *Social Studies of Science*, *30*(2), 225–263.

deNoyelles, A., & Raider-Roth, M. (2015). Being an 'agent provocateur': Utilising online spaces for teacher professional development in virtual simulation games. *Technology, Pedagogy and Education*, *25*(3), 337–353.

Edwards, C. P. (2000). Children's play in cross-cultural perspective: A new look at the Six Cultures study. *Cross-Cultural Research*, *34*(4), 318–338.

Escobar, A. (2008). *Territories of difference: Place, movements, life, redes*. Durham, NC: Duke University Press.

Facer, K. (2011). *Learning futures: Education, technology, and social change*. New York, NY: Routledge.

Ferreira, P. (2015). *Play as freedom: Implications for ICT4D* (Unpublished doctoral dissertation). KTH Royal Institute of Technology, Stockholm, Sweden.
Fisher, J. (2016a). Toward a political economic framework for analyzing Digital Development Games: A case study of three games for Africa. *Communication, Culture & Critique*, *9*(1), 30–48.
Fisher, J. (2016b). Digital development and international aid: Are games changing the world? In H. Gangadharbatla & D. Davis (Eds.), *Emerging research and trends in gamification* (pp. 276–303). Hershey, PA: IGI Global.
Fishman, B., Riconscente, M., Snider, R., Tsai, T., & Plass, J. (2014). *Empowering educators: Supporting student progress in the classroom with digital games*. Ann Arbor, MI: University of Michigan.
Flanagan, M. (2009). *Critical play: Radical game design*. Cambridge, MA: The MIT Press.
Gee, J. P. (2007). *What video games have to teach us about learning and literacy*. New York, NY: Palgrave Macmillan.
Gutiérrez, K. D. (2008). Developing sociocultural literacy in the third space. *Reading Research Quarterly*, *43*(2), 148–164.
Gutiérrez, K. D., & Penuel, W. R. (2014). Relevance to practice as a criterion for rigor. *Educational Researcher*, *43*(1), 19–23.
Hoadley, C., & Cox, C. (2009). What is design knowledge and how do we teach it? In C. DiGiano, S. Goldman, & M. Chorost (Eds.), *Educating learning technology designers: Guiding and inspiring creators of innovative educational tools* (pp. 19–35). New York, NY: Routledge.
Holden, J. (2013). Playful possibilities for assessment: Fluffy ducks and the queen's gambit. *The Scholarship of Teaching*, *6*(1), 4–8.
Holden, J. (2016). Good game: On the limitations of puzzles and possibilities for gameful learning. In C. Williams-Pierce (Ed.), *Teacher pioneers: Visions from the edge of the map*. Pittsburgh, PA: ETC Press.
Holden, J., Kupperman, J., Dorfman, A., Saunders, T., Pratt, A., & MacKay, P. (2014). Gameful learning as a way of being. *International Journal of Learning Technology*, *9*(2), 181–201.
Holden, J., Stanzler, J., Fahy, M., & Kupperman, J. (2015). Priorities and pathways for civic caucusing: The Michigan Student Caucus. In *Civic media project*. Retrieved from http://civicmediaproject.org/works/civic-media-project/civiccaucusingmichigan
Hussain, S., & Al Barwani, T. (2015). Sustainable development at universities in the Sultanate of Oman: The interesting case of Sultan Qaboos University (SQU). In W. Filho (Ed.), *Transformative approaches to sustainable development at universities: Working across disciplines* (pp. 151–163). New York, NY: Springer International Publishing.
Hwang, G. J., & Wu, P. H. (2012). Advancements and trends in digital game-based learning research: a review of publications in selected journals from 2001 to 2010. *British Journal of Educational Technology*, *43*(1), E6–E10.
Johri, A. (2011). The socio-materiality of learning practices and implications for the field of learning technology. *Research in Learning Technology*, *19*(3), 207–217.
Kam, M., Agarwal, A., Kumar, A., Lal, S., Mathur, A., Tewari, A., & Canny, J. (2008, February). Designing e-learning games for rural children in India: A format for balancing learning with fun. In *Proceedings of the 7th ACM conference on Designing interactive systems* (pp. 58–67). Cape Town, South Africa: ACM.
Kam, M., Kumar, A., Jain, S., Mathur, A., & Canny, J. (2009, April). Improving literacy in rural India: Cellphone games in an after-school program. In *Information and Communication Technologies and Development (ICTD), 2009 International Conference on* (pp. 139–149). Doha, Qatar: IEEE.
Kangas, M. (2010). Creative and playful learning: Learning through game co-creation and games in a playful learning environment. *Thinking Skills and Creativity*, *5*(1), 1–15.
Killham, J., Tyler, S. P., Veneble, A., & Raider-Roth, M. (2014). Mentoring in an online simulation: shaping pre-service teachers for tomorrow's roles'. *Teaching and Learning: The Journal of Natural Inquiry and Reflective Practice*, *28*(2), 62–79.
Kolko, B. E., & Putnam, C. (2009, April). Computer games in the developing world: The value of non-instrumental engagement with ICTs, or taking play seriously. In Information and

Communication Technologies and Development (ICTD), *2009 International Conference on* (pp. 46–55). IEEE.

Kumpulainen, K., & Sefton-Green, J. (2014). What is connected learning and how to research it? *International Journal of Learning and Media*, *4*(2), 7–18.

Kupperman, J. (2010). *DIGW in Kingston, Jamaica*. Retrieved from https://www.hastac.org/blogs/jkupp/2010/05/11/digw-kingston-jamaica

Kupperman, J., Fahy, M., Goodman, F., Hapgood, S., Stanzler, J., & Weisserman, G. (2011). It matters because it's a game: Serious games and serious players. *International Journal of Learning and Media*, *2*(4), 21–30.

Kupperman, J., Robertson, B., & Baglin, S. (2010). DevInfo GameWorks: Supporting inquiry-based game design. In K. Gomez, L. Lyons, & J. Radinsky, (Eds.), *Learning in the disciplines: Proceedings of the 9th International Conference of the Learning Sciences* (ICLS 2010, Vol. 2, pp. 296–297). Chicago, IL: International Society of the Learning Sciences.

Kupperman, J., Stanzler, J., Fahy, M., & Hapgood, S. (2007). The Grasshopper: Games, school, and the benefits of inefficiency. *International Journal of Learning*, *13*(9), 161–168.

Lave, J., & Wenger, E. (1991). *Situated learning: Legitimate peripheral participation*. Cambridge, UK: Cambridge University Press.

Lemke, J., Lecusay, R., Cole, M., & Michalchik, V. (2015). *Documenting and assessing learning in informal and media-rich environments*. Cambridge, MA: The MIT Press.

López Turley, R. N., & Stevens, C. (2015). Lessons from a school district-university research partnership: The Houston Education Research Consortium. *Educational Evaluation and Policy Analysis*, *37*(1S), 6S–15S.

Moje, E. B., Ciechanowski, K. M., Kramer, K., Ellis, L., Carrillo, R., & Collazo, T. (2004). Working toward third space in content area literacy: An examination of everyday funds of knowledge and Discourse. *Reading Research Quarterly*, *39*, 38–70.

Pal, J., Pawar, U. S., Brewer, E. A., & Toyama, K. (2006, May). The case for multi-user design for computer aided learning in developing regions. In *Proceedings of the 15th international conference on World Wide Web* (pp. 781–789). New York, NY: ACM.

Paley, V. G. (1992). *You can't say you can't play*. Cambridge, MA: Harvard University Press.

Patra, R., Pal, J., & Nedevschi, S. (2009). ICTD state of the union: Where have we reached and where are we headed. In *Information and Communication Technologies and Development (ICTD), 2009 International Conference on* (pp. 357–366). Doha, Qatar: IEEE.

Peppler, K., & Kafai, Y. B. (2010). Gaming fluencies: Pathways into a participatory culture in a community design studio. *International Journal of Learning and Media*, *1*(4), 1–14.

Resta, P., & Laferrière, T. (2015). Digital equity and intercultural education. *Education and Information Technologies*, *20*(4), 743–756.

Sachs, W. (2010). Globalization, convergence and the Euro-Atlantic development model. In M. Redclift & G. Woodgate (Eds.), *The international handbook of environmental sociology* (2nd ed., pp. 262–275). Northampton, MA: Edward Elgar Publishing.

Salen, K. (Ed.). (2008). *The ecology of games: Connecting youth, games, and learning*. Cambridge, MA: The MIT Press.

Salen, K., & Zimmerman, E. (2004). *Rules of play: Game design fundamentals*. Cambridge, MA: MIT press.

Salen Tekinbas, K. (2014). Learning to pivot: A play on possibility. In S. Walz & S. Deterding (Eds.), *The gameful world: Approaches, issues, applications* (pp. 625–628). Cambridge, MA: MIT Press.

Salen Tekinbas, K., Gresalfi, M., Peppler, K., & Santo, R. (2014). *Gaming the system: Designing with Gamestar Mechanic*. Cambridge, MA: The MIT Press.

Schmitz, B., Felicia, P., & Bignami, F. (2015). Acceptance and use of game-based learning in vocational education and training: An international survey. *International Journal of Game-Based Learning*, *5*(4), 50–63.

Sicart, M. (2014). *Play matters*. Cambridge, MA: The MIT Press.

Slater, M. (1999). Measuring presence: A response to the Witmer and Singer presence questionnaire. *Presence Teleoperators and Virtual Environments*, *8*(5), 560–565.

Smith, P. (2010). *Children at play*. West Sussex, UK: Wiley-Blackwell.

Soja, E. (1996). *Third space*. Cambridge, MA: Blackwell.
Souto-Manning, M. (2011). Playing with power and privilege: Theatre games in teacher education. *Teaching and Teacher Education, 27*(6), 997–1007.
Squire, K. (2010). From information to experience: Place-based augmented reality games as a model for learning in a globally networked society. *Teachers College Record, 112*(10), 2565–2602.
Squire, K. (2011). *Video games and learning: Teaching and participatory culture in the digital age*. New York, NY: Teachers College Press.
Steinkuehler, C., Squire, K., & Barab, S. (Eds.). (2012). *Games, learning, and society: Learning and meaning in the digital age*. Cambridge, UK: Cambridge University Press.
Stevens, R., Satwicz, T., & McCarthy, L. (2008). In-game, in-room, in-world: Reconnecting video game play to the rest of kids' lives. In K. Salen (Ed.), *The ecology of games: Connecting youth, games, and learning* (pp. 41–66). Cambridge, MA: The MIT Press.
United Nations Development Programme. (2010). *The real wealth of nations: Pathways to human development*. New York, NY: Palgrave Macmillan.
Williams-Pierce, C. (2016). *Teacher pioneers: Visions from the edge of the map*. Pittsburgh, PA: ETC Press.
Wouters, P., & Oostendorp, H. V. (2013). A meta-analytic review of the role of instructional support in game-based learning. *Computers & Education, 60*(1), 412–425.
Young, M. F., Slota, S., Cutter, A. B., Jalette, G., Mullin, G., Lai, B., … Yukhymenko, M. (2012). Our princess is in another castle: A review of trends in serious gaming for education. *Review of Educational Research, 82*(1), 61–89.
Zhang, J., Yang, J., Chang, M., & Chang, T. (2016). Towards a critical understanding of the best practices of ICT in K-12 education in global context. In J. Zhang, J. Yang, M. Chang, & T. Chang (Eds.), *ICT in education in global context* (pp. 1–17). Singapore: Springer Singapore.

Author Information

Jeremiah H. Kalir
School of Education and Human Development, University of Colorado Denver, Campus Box 106, 1380 Lawrence Street, Suite 645, Denver, CO 80204, USA

Complete mailing address: School of Education and Human Development, University of Colorado Denver, Campus Box 106, 1380 Lawrence Street, Suite 645, Denver, CO 80204, USA
Telephone number: (303) 315-0034
Email address: remi.kalir@ucdenver.edu
Website: www.remikalir.com

Remi Kalir, is a learning scientist and teacher educator whose interests and design-based research concerns learning across settings, educators' design of games, and mobile learning. As Assistant Professor of Information and Learning Technologies at the University of Colorado Denver School of Education and Human Development, Remi is a Creative Research Collaborative Fellow and a member of the University of Colorado President's Teaching and Learning Collaborative. His work regularly establishes research-practice partnerships with schools and educators, leading to the collaborative design of technology-mediated learning for interdisciplinary inquiry and civic engagement across school, online, and community settings.

Michael Fahy
University of Michigan, School of Education, University of Michigan, 610 E. University Ave., Ann Arbor, Michigan 48109

Complete mailing address: School of Education, University of Michigan, 610 E. University Ave., Room 4007, Ann Arbor, MI 48109-4089
Telephone number: (734) 646-4089
Email address: michfahy@umich.edu

Michael Fahy is a Lecturer in Educational Studies and a member of the University of Michigan's Interactive Communications & Simulations Group (ICS) and the Institute for Innovation in Education (iiE). With a background in Anthropology (Ph.D. University of Michigan, 1998) and research in the Middle East and North Africa, his focal interests are conceptions of education and its forms of practice within cultural idioms, and the application of educational technology to foster mutual understanding across geopolitical and cultural boundaries.

Jeff Kupperman
University of Michigan-Flint, 303 E. Kearsley St., Flint, MI 48502

Complete mailing address: School of Education & Human Services, 430 French Hall, 303 E. Kearsley St., Flint, MI 48502
Telephone number: (734) 604-6117
Email address: jkupp@umich.edu
Website: www.jkupp.com

Jeff Kupperman is an Associate Professor of Education and a core member of the University of Michigan's Interactive Communications & Simulations (ICS) group and the Institute for Innovation in Education (iiE). A major focus of his work involves the development of technology-mediated programs that allow K-12 and university students to push the boundaries of play, communication, social action, and culture.

Farrah M. Schiff
Interactive Communications & Simulations, School of Education, University of Michigan, 610 E. University Ave., Ann Arbor, Michigan 48109

Complete mailing address: School of Education, University of Michigan, 610 E. University Ave., Ann Arbor, Michigan 48109
Telephone number: (607) 651-0765
Email address: fmschiff@gmail.com

Farrah Schiff earned her Master's degree in Educational Studies from the University of Michigan in 2010. Since then, she has been involved with various Interactive Communications & Simulations (ICS) projects, most frequently as a mentor and a project manager for the *Place out of Time* simulation.

Jeff Stanzler
School of Education, University of Michigan, 610 E. University Ave., Ann Arbor, Michigan 48109

Complete mailing address: School of Education, University of Michigan, 610 E. University Ave., Room 4007, Ann Arbor, MI 48109-4089
Telephone number: 734 (763-5950)
Email address: stanz@umich.edu
Website: www.ics.soe.umich.edu

Jeff Stanzler is a faculty member at the University of Michigan School of Education, working in teacher education and digital media, and directing the Interactive Communications & Simulations (ICS) group, which facilitates mentored simulations and writing projects for K-12 students worldwide. Jeff is on the faculty of the Master of Arts in Educational Studies with Secondary Teacher Certification (MAC) Program and the Michigan Community Scholars Program and is a faculty affiliate of the Center for Middle Eastern and North African Studies, and the Program in Intergroup Relations.

Chapter 8
Impatience as a Virtue: Addressing Persistent ICT-in-Education Challenges in Small Developing Countries

Laura Hosman

Abstract This chapter details one scholar's experience being a practitioner in her area of expertise: technology for schools in developing world contexts. After being invited to design a solar powering system to charge donated laptops in Haitian schools, she began working with students in hands-on, project-based courses to design and deploy ICT4D-related technologies in the field. Once on the ground, she and her students discovered that the challenges and failures of playing such an active role can be many, but the speed and intensity with which we learn is much greater than from more traditional types of classroom-based or 'book learning'. After shifting geographic focus to Pacific Island schools, the author and her students designed, developed, and deployed multiple new technologies, including a solar computer lab in a box and a solar digital library. Throughout this venture, unexpected developments continually arose, which have led to the changing of project partners, improvement of the relevance and usefulness of technology being developed, and an expansion of the project to multiple small island-nations. Through a narrative case study of experiential learning, both in-the-classroom and in-the-field, this chapter details the challenges and lessons learned from the scholar-practitioner's experience.

Keywords Experiential Education • Learning-by-Doing • Project-Based Courses • Scholar-practitioner

---

L. Hosman (✉)
Arizona State University, Interdisciplinary B256K,
1120 S. Cady Mall, P.O. Box 875603, Tempe, AZ 85287-5603, USA
e-mail: laurahosman@gmail.com; https://ict4dviewsfromthefield.wordpress.com

I.A. Lubin (ed.), *ICT-Supported Innovations in Small Countries and Developing Regions*, Educational Communications and Technology: Issues and Innovations,
DOI 10.1007/978-3-319-67657-9_8

1 Introduction

This chapter documents both the difficulties faced and progress made by a scholar who, after years of researching the issue of introducing ICTs into remote, rural, developing world schools, decided to take a hands-on, action-oriented approach to the challenge and become a scholar-practitioner. The challenges of taking this unconventional approach are many and present themselves at multiple levels. At the personal level, there is the challenge of doing work one finds meaningful and impactful while bravely venturing into new realms of activity in which one has little or no experience. This must be done while remembering one's obligations to the scientific endeavor of carrying out research and finding funding to accomplish the work. At the pedagogical level, the challenge is providing students with an authentic, experiential, learn-by-doing experience that actually benefits a partner community on the other side of the globe. At the level of scholarly endeavor, the challenge arises out of trying to connect this teaching, research, and learning into the application of new knowledge, both tangibly (in the field) and scientifically (in publications), and establishing the feedback loops so that this new knowledge can be applied meaningfully.

Furthermore, being a scholar-practitioner requires focusing on all of these challenges: the personal, pedagogical, research, ecological (local and global), and developmental (reflection and learning), *simultaneously*. Many of the challenges faced by the scholar-practitioner lie outside of her/his control, yet the keys to making improvements is comprehending the nature of such challenges and learning from setbacks while still seeking pathways forward.

The context in which my activities and research take place are resource-constrained schools in the developing world. There is a drive to introduce information and communication technologies (ICT) into schools that appears to be increasingly universal in nature, motivated at least in part by the belief that being able to use ICT is necessary for participation in the global information, knowledge, digital, and technology-based economies of the future (Kozma & Voogt, 2003). Yet for schools in remote, rural, isolated, and resource-constrained locations, especially in developing regions and small states, the difficulties associated with introducing ICT are manifold. Just some of the attendant challenges include lack of electricity and/or Internet connectivity, no funding to train teachers on the use of ICTs, no relevant, localized digital content for teaching, and no locals with the skills to train teachers or to create digitized educational content or curricula (Williams, 2009). Addressing any one of these challenges would be difficult, but for technology to be adopted successfully, it would seem that schools may need to address all of these issues simultaneously. As a result, schools and departments of education have not known where to start in terms of introducing education-relevant technologies. These are the circumstances within which my activities and research take place. The questions driving my work include the following: What does appropriate, or even usable, technology look like in schools where it has never been used before? What are the best practices, or even minimum requirements, to help technologies be adopted and used fruitfully to improve educational opportunities and outcomes? Yet, the scholar-practitioner must add the following questions: How does one address these issues in a hands-on way while still developing as researcher, teacher, and scholar? How can such work be done in a manner that is reflexive and empowering to oneself, one's students, and to in-field partners?

This chapter, therefore, addresses the contemporary struggles of being an effective scholar-practitioner in one's field of expertise. Taking this unconventional action-oriented approach to how one understands one's topic of study has led to multiple unexpected insights, pivot points, and lessons-learned. Involving students in the process in a hands-on way, from the classroom to the field, has allowed for transformative educational and life experiences to take place for the faculty member and students alike. The challenges have been many, but the experiences are always extremely rewarding.

The next section provides an overview of the methodology used in the chapter: analytic autoethnography. It continues with a description of the fieldwork that is divided into three categories that mirror the different technologies and map to the geographic regions of work. It continues with a discussion of pedagogical challenges and lessons learned from in-the-field experiences as I, the scholar-practitioner, developed and revised my approaches to technologies as well as to my teaching and research over time.

2 Methodology

This section defines and discusses the methods used in this chapter as well as the data collection and triangulation process. The chapter employs a multi-sited, analytic autoethnographic approach to present and examine an in-depth case study. Brewer (2000), using a traditional definition, describes ethnography as a data collection method meant to capture the social meanings of people and activities taking place in naturally occurring settings, commonly referred to as "the field." The goal is to collect data in such a way that the researcher imposes a minimal amount of his/her own bias on the data (Brewer, 2000, p. 10).

By utilizing ethnographic methods and adopting a narrative voice in this work, I heeded a call for increased use of this marginalized yet invaluable methodological approach (Bayard de Volo & Schatz, 2004;Patterson & Renwick Monroe, 1998 ; Weeden, 2010). This method is employed with an eye toward (1) making sense of the disparate facts that construct our reality, (2) gaining broader and more generalizable insight from the case at hand, (3) interpreting and understanding the realities around us, and (4) contributing to theory development and testing (Patterson & Renwick Monroe, 1998; Weeden, 2010).

Yet, in order to capture the richness and complexity of both the research and the experience, I added two specifications to the methodological concept of ethnography – the research was multi-sited and autoethnographic. This work was necessarily multi-sited in that the research described herein took place in multiple locations (on and around an American campus, and in Haiti, Micronesia, Samoa, and Vanuatu).

Pierides (2010) argues that multi-sited ethnography presents a more realistic way to portray the political economy of our globalizing world.[1] Multi-sited ethnography is inherently comparative in nature, as the researcher analyzes the relationships and associations between phenomena observed across multiple real-world sites of investigation (Marcus, 1995).

When a researcher is intimately involved in shaping the process and events under analysis, as was the case in the present study, the researcher is both participant and observer. Such activity aligns with what Anderson (2006, p. 378) has proposed as "analytic autoethnography." He uses the term to characterize qualitative research that satisfies five criteria, all of which apply in the case at hand. The researcher has the following characteristics:

1. Is a full member of the research group.
2. Is aware of the fact that she/he forms part of the process under investigation but is also in part formed by those processes as she/he co-creates the reality she/he records and analyzes.
3. Is a highly visible social actor within the text.
4. Is in dialogue with informants and participants beyond her/himself.
5. Is committed to gaining insight and developing theoretical understandings of broader social phenomena.

When the researcher is intimately engaged in the process and events under examination and is the sole source of data, description, and analysis, the possibility arises for potentially diminished criticality, neutrality, or comprehensiveness. In order to avoid and/or ameliorate this, triangulation of data took place among multiple actors. During the trips to the field, student-participants made daily video travelogues/diaries to assist with their own reflection on what they experienced. These travelogues combined with active discussions between students, project partners, and myself to confirm the data and ensured it completeness – the two main purposes of triangulation (Houghton, Casey, Shaw, & Murphy, 2013). Ongoing dialogue with all participants also served as a valuable source for data collection and verification.

In addition to enabling the triangulation of data, there are other salient reasons to involve students in the autoethnographic research process. Central to the benefits realized through the use of ethnographic research, Bayard de Volo and Schatz (2004) assert that,

> Through close encounters...young scholars gain sensibilities that potentially inform their whole careers. The consequences of good or bad governance become concretely connected to real-life beneficiaries or victims of state policies. The moral issues at stake in studies of poverty, inequality, or political repression become quite palpable. ... early experience of the sort that ethnography provides can be all the more important. (p. 270).

[1] In an interesting and coincidental footnote, George Marcus, often seen as the originator of the idea of multi-sited ethnography, arrived at this concept springing from the disconnect he felt when trying to apply Immanuel Wallerstein's World Systems Theory to the realities on the ground as Marcus carried out ethnographic research in Tonga, another Pacific Island (Pierides, 2010: 185).

In other words, ethnographic research provides a teachable moment par excellence for students who are participating or have just participated in the hands-on, action-oriented, and learn-by-doing courses described herein. What is more, giving students the opportunity to record, reflect upon, and process their feelings, thoughts, and behaviors helped them think more deeply about both their own beliefs and the beliefs of the communities with which they interacted (Tomlinson-Clarke & Clarke, 2010). As a result, students reported becoming more articulate and in tune with their own beliefs, learning more about the communities with which they interacted, making personal connections within those communities, and feeling a greater sense of immersion (Maher, 2003).

In addition to ethnography/analytic autoethnography, I employed the use of the case study method. Case study research is useful for the study of a phenomenon in its natural context and is a preferred method when examining contemporary events where the author is a direct observer (Yin, 2003). In this case, I set out to study the introduction and uptake of ICT in remote rural schools. Yet, the very process of self-reflection combined with the goals of becoming a scholar-practitioner, while also offering meaningful educational experience to students, itself allowed for additional data collection and further analyses informing the case study described in this chapter. An additional goal of employing case study is to attempt to generalize from specific findings and contribute to theory development (Gerring, 2004).

A frequently raised concern with case study methodology is the lack of rigor. However, in this case, this concern was addressed through the triangulation of data. In fact, the opportunity to triangulate data from different sources is a major strength of case study research (Yin, 2003). In both of these respects – in using triangulation to improve rigor and increase understanding, and in contributing to theory development and generalization of insights – case study research has strong synergies with ethnography/analytic autoethnography as defined above (Bayard de Volo & Schatz, 2004; Patterson & Renwick Monroe, 1998; Wall, 2006).

The following section presents a description of the case under analysis that took place both on a US campus and abroad utilizing an ethnographic/analytic autoethnographic approach. The subsequent section discusses some of the benefits, challenges, and lessons learned about constructivist experiential education and service learning through this experience.

3 Scholar-to-Practitioner

My desire to pursue experiential learning and carry out hands-on projects in my field(s) of study sprang from a deep dissatisfaction with the idea that writing papers would be my greatest impact and the one by which I would be measured, evaluated, and judged as an academic. I realize, of course, that we do make an impact in the classroom when we teach. We reach our students and we can change the course of their lives, but if this happens at all, it tends more frequently to happen far after we are no longer in our students' lives. Furthermore, it is the rare person who takes the time to reflect and reach out with a word of thanks or even an update. I am among the guiltiest of these people, as far as reaching out, giving updates, or saying thanks, so I am always astonished when a student does take the time to do any one of these things.

Thus, I was searching for a way to have a more meaningful impact in the real world in my area of specialization while remaining an academic. My area of academic specialization is information communications technology for development (ICT4D). Within this field, I have a particular passion for understanding why it is the case that when technology is brought to schools where it has never been used before, these initiatives so often fail – and what can be done to improve technological uptake, adoption, and the resulting educational experiences of both teachers and students.

When, as a post-doc, I finally had the opportunity to carry out fieldwork, I was overwhelmed by how much faster I learned and by how so many things that matter – big and small, ranging from cultural to infrastructural realities – were left out of the academic literature. Some of this is out of necessity: adhering to an academic tone, retaining focus, and keeping to a publishing word limit, all while providing evidence to address a research question, may leave little room for anecdotal observations. Yet it is the rich nuances and complex realities that often hold the key to a more profound understanding (of what may or may not lead to improvements and why) and the opportunity for learning. Thus, I set out to address my research questions head-on; by taking an active, hands-on approach to working with teachers, schools, ministries and departments of education, administrators, and practitioners. Over the course of the next few years, I would form partnerships between multiple organizations; carry out in-field technology deployments; interview and discuss with teachers about their technology beliefs, fears, and actual usage; and even eventually spearhead the creation of new technologies for in-school implementations, based on what I had experienced and learned from my years of experience in the field. In other words, I became a scholar-practitioner.

Since hands-on fieldwork proved to be transformational for me, I believed it would prove similarly life-changing for my students. So, I also set about creating project-based courses that would allow students the opportunity to become involved as well.

3.1 Hands-on, Project-Based Courses Focused on Real-World Challenges

All of the courses alluded to in this chapter share many similar characteristics, regardless of the specific university where they were offered, the project being undertaken, or the area of geographic partnership and activity. Each class was project-focused, team-based, and comprised of students from multiple majors. Students from any major are always welcome to apply to be in the classes, and although at times certain majors were specifically sought (such as software engineering), the majority of students have always been admitted based on their demonstrated level of interest in being actively involved in the given project.

In all cases, at the beginning of the term, students are presented with one or multiple real-world challenges they are expected to address and the knowledge that our in-field partner is counting on them to deliver some type of solution. Students need to agree on what the deliverable will be and then have the balance of the semester or quarter to figure out how they will accomplish this goal. Every class forms subteams, based on interests and areas of expertise, to tackle a specific aspect of realizing the deliverable or otherwise advancing the project. These teams could be designing and building prototypes, carrying out research on local culture, politics, history, etc., testing existing technology, creating teaching modules, making surveys, setting up websites, or fundraising. This is not an exhaustive list, and new teams are formed with each successive class in order to address the needs specific to that semester's challenge. Outside experts are frequently consulted and invited to mentor the students, and the class size generally ranges from 10–16 students.

The teams are also expected to provide regular status updates as well as presentations about the project and their role in its realization to outside experts and to our partners in the field. In effect, the courses are designed to mimic a team-based work environment that students are increasingly likely to encounter in a professional setting after they graduate. When active learning and a hands-on approach are used in instruction, they inspire students' personal interest in the subject matter, lead to intrinsic motivation, and boost overall comprehension of the subject matter (Kapp, 2009; Peterson, 2009). Moreover, the practical application and simulation of work experience make students more attractive to the competitive workforce (Marlin-Bennett, 2002).

At the conclusion of nearly all of these courses, the faculty member and (some) students are able to travel to the fieldwork site and see through to fruition on-site the project work they have been carrying out while on campus – working in tandem with our in-field partners and having a far richer learning experience than staying on campus. This is evidenced in students' own feedback and journals. The students' on-site journaling and post-course feedback are used to gauge learning outcomes and promote the kind of deep reflection described in the methods section above. It is beyond the scope of this chapter to discuss this topic at great length, however, all students participating in both on-campus and off-campus portions of these courses are asked to complete a survey intended to gauge their experience vis-à-vis multidisciplinary teamwork, active learning, cultural competency, professional plans, and (if relevant) their travel experience and how it might affect all of the previous categories. Additionally, all students who travel are asked to keep journals or video-journals and to become autoethnographers, knowing that their self-reflections may be used for scholarly purposes. The courses thus feature innovative pedagogical approaches, such as experiential service-learning, multidisciplinary collaboration, and international immersion and have established the mechanism for measuring the self-reported impact on the students.

The timeline in Table 8.1 is intended to provide a frame of reference for the reader since there is an overlap of both (home) universities and project/partner locations. Yet, it is also important to note that the format of the courses as well as students' ability to travel and carry out fieldwork have remained the same throughout.

Table 8.1 Timeline of project activities and locations

Year	2010	2011	2012	2013	2014	2015	2016
University	Illinois Institute of Technology					California Polytechnic State University	
Partner location	Haiti			Federated States of Micronesia (FSM)		Vanuatu & FSM	
Project	Solar system for laptops & solar powered internet connectivity			Solar computer lab in a box		Solar digital library	

4 Solar Powering for Laptops and Solar Long-Distance WiFi in Haiti

My first opportunity to become a scholar-practitioner came in late 2009. I was presented with the opportunity to work on developing a solar-powering system to provide charging capability for laptops that had been donated to the Ministry of Education in Haiti. The laptops came from the One Laptop per Child (OLPC) Foundation, and to carry out this endeavor, Haiti's Ministry of Education had formed a quasi-independent organization called OLPC-Haiti. The director of OLPC-Haiti approached a colleague of mine about designing a solar powering system so that schools receiving these laptops would be able to charge them. My colleague approached me about becoming involved and proposed that I involve my students in the endeavor. I was able to offer the course as a service-learning, project-based, hands-on course, as described in the section above. Elsewhere, I have described this project in greater detail (Hosman, 2014), and examined the partnerships formed as part of this project as well as the lessons learned over the course of this two-year partnered project.

We accomplished quite a bit in the first three semesters that my students worked on these various projects in Haiti. We managed to design and deploy a 2.4 KW solar system at a primary school in Lascahobas, Haiti. The purpose-built system met the Ministry's request for the system to be easily replicable and more affordable than the price quotes they had received from local solar providers (whom we also worked with). Additionally, since our on-the-ground partner offered no funding toward the solar system or our travel, we raised over $40,000 to purchase the equipment (which we did through local partners), to travel to Haiti, and to deploy the system. Figure 8.1 is a picture of my students, colleague, and volunteers from a Haitian-based solar powering company installing the solar panels on the roof of the primary school where we worked.

After the initial solar installation took place (in 2011), we returned 6 months later to the same primary school and set up a solar-powered, long-distance, point-to-point WiFi Internet connection. On both of these deployment trips, we trained and worked with local Haitians, took on Haitian university students as interns, held meetings with local parents, and conducted technology-training workshops with teachers at the school. We tried to do everything that I knew was a *best practice* from my

Fig. 8.1 Setting up the solar system in Haiti

research. Yet best practices were quickly forgotten when logistical considerations arose such as the occurrence of earthquakes, having insufficient funds for travel, or the provost attempting to cancel our trip with only a month's advance notice. I learned that emergencies and other urgent challenges inevitably arise and you must address them as best you can. The only and best preparation is having laid the solid groundwork and having reliable and trustworthy partners.

Yet, one single aspect of the project has had the greatest impact on me, on my research, and on my motivation to continue doing this work. It is the self-same reason my students and I were invited to contribute to it in the first place. Haiti's Ministry of Education had received this donation of 11,000 laptops matched with a $25 Million (USD) budget from the Inter-American Development Bank (IDB) to enable the rollout of the laptops. Yet there was no funding allocated for electricity or infrastructure for the schools (IDB, 2007). This was a major oversight, as nearly 95% of Haiti's primary schools lack electricity.

This experience has helped me comprehend that it is a part of human nature to take for granted anything that has become commonplace in our lives: we notice the novel and the troublesome, and we overlook the things that have become normal to us, even as we may forget that these things may not be so normal to others. This happens all over the world, even on the part of well-intentioned, intelligent people. In fact, just a few weeks into working on the project, my very first group of students was required to make a presentation about our work to the other professors and students in this project-based program. During the question-and-answer time, the very first question my students received came from an engineering professor, who wanted to know why we were working on this project at all and why there was a need to bother with solar energy: 'Why can't they just plug the laptops into the wall?'

Taking things for granted is problematic when technology or development interventions are conceived and designed in one place, but the intended beneficiaries/recipients live in a different reality altogether. There can be an inability to consider and empathize with cultural and social differences when one does not have the requisite experience. When we are describing the interaction between those in wealthy parts of the world attempting to assist those in poor parts of the world, the burden of comprehending the realities facing those on the ground must fall to those with the resources, that is, those in the materially wealthy world. Unfortunately, the rich benefactors also have the least to lose by failing to make these considerations. When asked about the $25 million dollar laptop project in Haiti, a high-ranking IDB officer's response was: 'We don't talk about that project.' To my ears, this was more or less the same as hearing: 'We don't learn from our failures because we don't need to.'

Because I find this unacceptable, I've focused on getting out into the field and trying to learn what the real challenges are on the ground and trying my best to remember not to take things for granted. In 2012, after the $25 Million from the IDB ran out and Haiti's Ministry of Education decided not to continue funding this laptop project with its own money, we lost our project partner. Only 3000 of the 11,000 laptops were ever distributed to schools due to the severe shortage of schools with the resources to charge and power the laptops.

5 New Project, New Location, New Incentives: The Solar Computer Lab in a Box (SCLIAB) in Chuuk, FSM

While we were still in Haiti, my students and I had seen firsthand how challenging it would be for the primary school's existing employees to maintain advanced solar technology, given that this was the first time they were introduced to anything like it. It had also taken multiple mentors and advisors for my team of mainly engineering majors to design and size the solar installation and the connectivity project. By contrast, solar powering or long-distance WiFi connectivity experts would rarely be available in rural schools like these nearly anywhere in the developing world. We had streamlined the technology but we would still need to simplify it by another order of magnitude if there was to be any hope of such technology being adopted, meaningfully used, and maintained by those with no prior exposure – in other words, by a typical teacher in a remote rural school.

With this in mind, the next challenge (in 2012) that I presented to my students was to develop a Solar Computer Lab in a Box (SCLIAB), which would be portable, could be set up and understood by a non-engineer (like myself), and would need as little maintenance as possible. And the students did it! Within 4 months, in a hands-on, team-oriented, project-focused class, they had designed and built multiple iterations of this SCLIAB. This pivot in terms of project and technology also marked a geographic shift of focus, from Haiti to the Federated States of Micronesia. We had met and identified a reliable local partner who was from Chuuk, one of Micronesia's four Federated States, and who was eager to get technology into schools there.

5.1 Overview of the Federated States of Micronesia, FSM

Located in the Northern Pacific, the Federated States of Micronesia's (FSM's) population of nearly 100,000 is spread over 600 inhabited islands, just north of the Equator. Though the country's total land mass is just over 700 square km, it covers a longitudinal distance of over 7200 square km. The low population density combined with infrastructural challenges associated with such extremely remote locations means that, except in the capital cities, infrastructural development remains at a low level. The majority of economic activity across Micronesia remains fishing and subsistence farming, and the per-capita income hovers around $300 USD (CIA, 2017).

Figure 8.2 is a map of Micronesia. There are four Federated States: Yap, Chuuk, Pohnpei, and Kosrae. Our project work has mainly taken place in Chuuk, which is the most populous state, as well as the poorest, in terms of per-capita GDP, and the

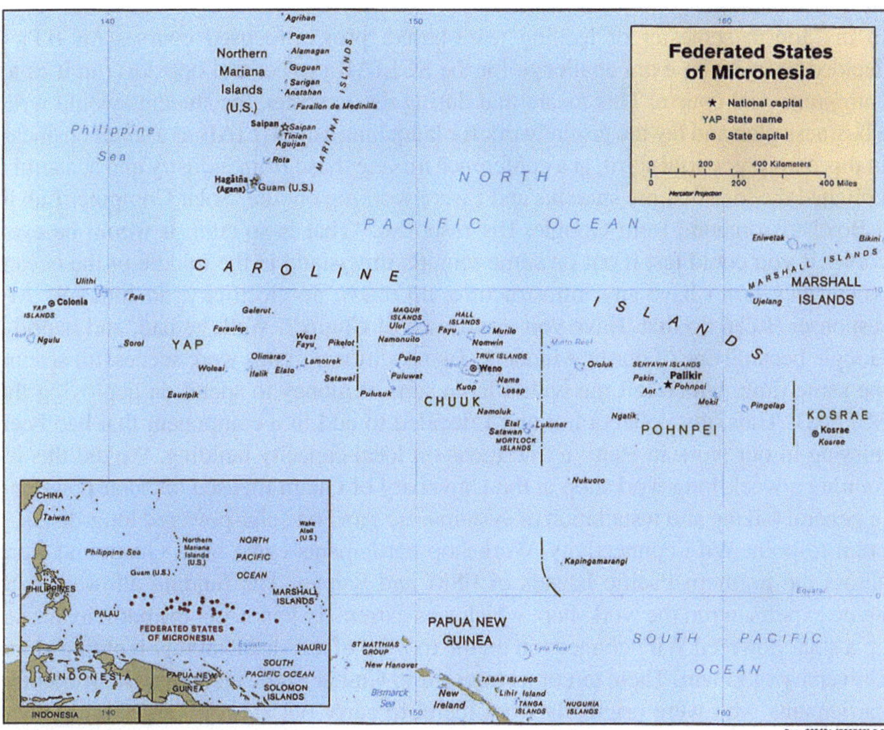

Fig. 8.2 Map of the Federated States of Micronesia (Source: Central Intelligence Agency (CIA) (1999))

most geographically dispersed, with by far the largest number of inhabited islands under its jurisdiction, which presents significant infrastructural challenges. The largest proportion of Chuuk's residents lives within the Chuuk Lagoon, an atoll with a protective reef, that holds 16 islands with significant populations of between 300 and 13,000. The largest island is also the capital, Weno. Most of these islands have no developed infrastructure, in terms of paved roads, sewer systems, electrical systems, and the like. Even the capital city only has a few paved roads (CIA, 2017).

Remote island populations face some of the greatest challenges in terms of being able to participate in the *development agenda*, as the materially wealthy countries have defined it. This is because the populations are so sparse, so spread out, and so difficult to reach. Yet, this is the very same reason that they stand to benefit so greatly from the advances that modern ICT could bring – the ability to communicate and to access information. I wanted to work in the Pacific Islands because I believed it to be the most challenging location on earth and that if we could address challenges there, we could address them anywhere.

5.2 Funding Opportunities, Partnering, Training, and Capacity Building

In my fourth semester of leading team-based, project-focused courses (at IIT), I decided to inject an extra challenge for the SCLIAB project and opted to run it as an entrepreneurial course. This meant that during the semester, the students would write a business plan and lay the groundwork for launching our SCLIAB as a small business. In this entrepreneurial spirit, at a conference mixing those from industry and academia, I pitched the idea that my students and I were working on: the 'Solar Computer Lab in a Box,' to someone from Google. His response: 'That is so cool. It would be even cooler if you could test it out on some remote, tiny island in the middle of the ocean, where they don't have any infrastructure: no roads, no electricity, no Internet.' My response: 'I can do that. Have you ever heard of Chuuk?' Well, he had, and happily, Google became one of our first funders. Two additional grants were successful around the same time, which left me with a large sum of money to spend on deploying the SCLIAB. Thus, my collaborator and I decided to add in a component that had been missing in our work in Haiti: a true focus on local capacity building. We did this by running a week-long workshop at the University of Guam focused on solar powering in general (sizing and installation of systems, etc.) and on solar-powered long-distance point-to-point WiFi connectivity. Workshop participants came from Guam and from across the northern Pacific Islands of FSM and Saipan. The funding allowed us to invite experts to run the workshop, which was extremely hands-on: the participants set up a solar powered WiFi connection on the rooftop of the sustainability building at the University of Guam. Then, ten of us traveled to Chuuk State, including the workshop participants who were originally from there, to carry out an even larger in-the-field

Fig. 8.3 Collage of the Solar Computer Lab in a Box, from in-box to assembled stages

deployment. Figure 8.3 illustrates the Solar Computer Lab in a Box in its box state, being assembled into a table, and finally, transformed into a computer lab.

As part of this large training and deployment, we successfully set up the Solar Computer Lab in a Box on Udot Island, in Chuuk State, FSM. We also, working with locals who had just been trained at the workshop, set up a solar-powered long-distance WiFi Internet connection across a 19 km span of the Chuuk lagoon. And it worked! The students I had brought with me from the US and the students we had more recently trained in Guam all got to contribute to the successful setup – a very rewarding experience.

Yet, although we were delighted to proclaim the SCLIAB's implementation a great success, we also faced numerous challenges. The first, I should have realized on-site: the deployment had taken place in August; only the principal at the school had been present for some modicum of training in how to use the technology. Our local partner told us that he and his colleagues would return to Udot Island and provide additional training to the teachers during the school year, but we later realized that this was fraught with political difficulties for our partners. Additional training was an expensive proposition that we had no budget for. Furthermore, there was concern that just being proficient at a given technology did not qualify our partners to be teachers of that technology. Our next challenge arose the week after we left: another local partner who was providing our Internet connection to the school and whom we understood would be relocating offices in about 8 months, actually relocated offices the following week. We lost our Internet connection at that school.

The two partners I mention above have in fact remained two of our strongest to this day (4 years later). I have certainly come to realize that the realities your local partners are facing on the ground are simply more complicated and multifaceted than you may realize, or than they may ever let on. More factors will always be beyond your control, and the most important thing may be to just keep moving forward, whenever this is possible.

5.3 Lessons (and Limitations) Learned

After our experience in Haiti, I had resolved to form the appropriate partnerships in order to avoid our efforts remaining a pilot, or one-off deployment. After demonstrating the proof-of-concept on Udot Island and hoping to expand to multiple island schools, I was eager to partner with the Chuuk State Department of Education (DoE). Yet in my first meeting with the DoE, it became clear that working with them might not be the best route forward, or even feasible at all, due to lack of interest, local capacities, and funds.

Over the next few months, however, another grant application came back successful. This one was to carry out site visits to six Chuuk Lagoon island schools, while training locals in how to do this, in order to set up six additional Solar Computer Labs with long-distance WiFI connectivity. While carrying out these site surveys, we returned to Udot Island to check on the SCLIAB. Eight months after the initial deployment, we found the computer lab covered with a sheet – never having been used. My team had, in fact, carried out a *worst practice*: deploy and depart, when we had initially set up the SCLIAB.

So now I learned how difficult it is to actually implement best practices when you are running a project and there are a million moving parts, all distracting you, and all depending on you. As the project coordinator of the deployment trip in August, I was in charge of the workshop in Guam; multi-site travel for all of the participants; car rentals, hotels, and airfare, including for my own students and for the experts; the equipment for all of the (multi-sited) deployments; managing all of the partnerships to make the whole thing work; and I was juggling three grants to pay for it all. So, with fundraising/grant writing and grant management, public relations, travel logistics, project logistics, equipment logistics, and partner-relationship maintenance, perhaps it is more surprising that everything managed to work out as well as it did. It was just a bit less shocking in the end that I had overlooked a best practice, sacrificed in the name of achieving the feasible.

Yet it was eye-opening for everyone – not just me. The teachers on the island were embarrassed that 8 months had passed and they had not used the computers. We were able to carry out an impromptu training with three of the teachers on that day, and happily, the computers were used after that. However, I realized that local teacher training would be the most difficult aspect of this work. There was no budget for it, no comprehension of its importance, and no locals in the entire state who could carry it out. I did not see a way forward, and it felt like hitting a brick wall.

6 Solar Digital Libraries and Working with the Peace Corps Across the Pacific Islands

Exactly during this time that I was struggling to uncover a way forward vis-à-vis teacher technology training, which was also a few months after we had set up the Solar Computer Lab in a Box on Udot Island, we were contacted via the project's website contact form by Melody, a US Peace Corps Volunteer. Melody was stationed on a neighboring island in the Chuuk Lagoon and was wondering whether we could set up such a computer lab on her island, Romanum. Thus began a years-long (2012–2015) process during which I came to see US Peace Corps volunteers as fantastic partners in this technology-in-the-schools endeavor.

One of the biggest challenges in successfully introducing technology into schools for the first time is the lack of teacher training – most particularly when teachers have never used such technology. The less accustomed to using technology (computers, tablets, the Internet, etc.), the less likely the chances for successful adoption, uptake, and meaningful use in the schools (Hosman & Cvetanoska, 2013; Mumtaz, 2000; Wozney, Venkatesh, & Abrami, 2006).

I have visited multiple technology deployments around the world, and the majority of them offered teachers, the vast majority of whom were using a laptop for the very first time, less than a week's worth of training, and then told them: Go forth and transform how you teach! Start now! You are fully trained! Meanwhile, a week was not even sufficient training on how to use the hardware, let alone how to change a mentality and mindset that has been cultivated perhaps since the teacher was a young student; a mindset that does not include using computers.

I was facing this too in Micronesia. As a foreigner and a full-time academic, I could not spend unlimited time in the field myself, carrying out training (and I am not actually trained or qualified to carry out such training in the first place!). Thus, I could not address the challenge by myself. And the situation on the ground appeared indomitable: there was no funding for teacher technology training at all. When there is teacher training (not related to technology), the practice was to bring all teachers together on the main island/capital city for a week-long training and then send them back to their home island or the island on which they teach. Micronesia was not alone: a colleague at the University of the South Pacific confirmed that this had been the form factor for teacher technology training across the Pacific Islands. This approach does not work because; (1) a week is not long enough to be transformative and to comprehend how to change everything you've known up to now; (2) there is no support for the teachers once they return to their home island/schools, in terms of hardware/technology, or in terms of release time to change lesson plans; and (3) there is often resistance from the local principals, other teachers, or powers-that-be, who may not have received similar training or who may not be project-champions for many competing reasons (Thomson, 2015). Thus, a double-challenge faces many island states with small populations: there are no, or very few, indigenous teacher-technology-trainers (the smaller the population, the more likely this is to be the case: it is the reality in FSM) and there is no funding set aside for bringing in permanent master-trainers from abroad.

Following a very quick Internet search, I learned that US Peace Corps volunteers across the Pacific Islands serve for 2 years. The majority are English literacy teachers or facilitators tasked with building the capacities and professionalization of primary- and middle-school level teachers using technology where possible. The more I read, the more I recognized that these volunteers could help address one of the greatest challenges of getting technology used effectively in the schools. The volunteers' seemingly complementary work was an outstanding way to address the challenge I faced with long-term teacher technology training. Additionally, the technology of a digital library was perfectly aligned with the Peace Corps mission and goals. It would not be a complete solution, by any means – it would not reach all schools in the Pacific Islands or even in any given state – but it would still be worthwhile to see whether progress could be made.

We initially began working with two Peace Corps Volunteers stationed in the Chuuk Lagoon. Back at home, my university students started to work on two separate tasks that seemed important. The first team created a repository of open-access educational information that could be provided to the schools in the absence of Internet connectivity. They then created a website so that the information could be found in a user-friendly format (instead of as a database) and then loaded all of this information and the website onto a server. The second team created a teacher-training manual focusing on 14 lessons that would walk a trainer through showing someone how to use a tablet for the first time. Following this semester, in August 2013, we returned to Chuuk and distributed these training manuals, tablets, and solar charging systems to these two volunteers. We had not yet determined how to make our content server user-friendly enough to go out to the islands so we installed it at our partner's cyber-café, the only one in the state of Chuuk, where it was broadcast over WiFi (freely) to anyone interested.

We learned a few months later that one of the two Volunteers had dropped out of the Peace Corps for personal reasons so we were working just with Melody. Melody had no Internet on her island, so I didn't hear from her until late November when she had traveled back to the main Island of Weno and wrote to us from the cyber-cafe. She wrote in a desperate tone, describing how her school's principal (who also happened to be her local 'father,' as the Volunteers stay with families) had found a reason every week since August to avoid the teacher technology training from taking place. Could she instead work with the 8th-grade students?

Of course, I said yes. My true goal was for the technology (the tablets) to be useful somehow for the school. Yet here was another obstacle for technology use in the schools and for teacher technology training that I could not have dreamed up from my desk chair.

6.1 *Measuring the Learning Impacts: Perspectives from the Field*

The happy ending to this portion of the story is that the 8th-graders that Melody taught received the highest score in the entire state, among public schools, on the English language national exam. We both found this out because I was contacted, out of the blue, by a consultant from Australia who was working on a report about the state of Education in Chuuk. He had found my blog post about the tablets going out to Melody and her island and wanted to know whether it was Melody or the technology that had led to these much-improved outcomes. Further, he pointed out that the students' math scores were not improved so that the tablets might not necessarily improve the students' numerical skills. Thus, how could the tablets be better used to improve both math and English language skills at all grade levels?

The tone of these questions struck me as completely missing the point. The consultant had absolutely no knowledge of how the tablets were used, by whom, or for what purpose. But there is an implicit assumption made that merely introducing technology should bring about better outcomes if we can just figure out the secret. I let him know that the secret was a good teacher, not the technology, but that I would let Melody give the final answer in her own words.

I contacted Melody over email and shared the good news with her about her 8th-graders' outstanding exam results (which she had not known prior to this) as well as relayed the consultant's questions. Her response as to the role of the tablets was that she had used them as follows: if the students showed up on time for school in the morning (when they had a different teacher) and showed up on time again after lunch (when she was their English teacher) and behaved during their time with her, then they were allowed to use the tablets for fun after class was over. Thus, she had used the tablets mainly as incentive and reward and not at all for educational purposes.

6.2 *Another Challenge, Another Lesson Learned, Another Opportunity*

Recall that in 2013 we visited Melody's island and carried out a site inspection (along with five other islands). The original goal of the trip had been to set up an intranet connecting all of the schools in the Chuuk Lagoon to the Capital city of Weno. When we set up the long-distance point-to-point solar powered WiFi connection at Udot Island in 2012, we discovered that the satellite Internet connection was slow (not to mention unreliable, as it was disconnected the following week!). By contrast, the connection that did not require connecting to the satellite – an island-to-island-only intranet – was extremely high-speed. We hypothesized that with an intranet we would provide the information repository we had been working on to the Department of Education so that they could broadcast it out over the intranet that we would help implement. As indicated above, our local partner had been broadcasting this information repository over WiFi from his cyber-café. We had already trained five locals twice, on separate visits, with the skills necessary to carry out this work. Training included setting up long-distance solar-powered point-to-point WiFi connections, tower safety, and carrying out site surveys with open source software and smartphones. We believed we had the local expertise and the buy-in from relevant local entities such as the DoE and FSM Telecommunication Corporation.

Then the reality of obtaining grants, setting up infrastructure, and of getting an agreement with the National Telecom office to use their tower set in. We waited for a year for the agreement with Telecom, one that they had suggested in the first place since we had trained some of their staff on our multiple training trips. We also set out to co-author grant proposals with our local partners to fund the proposed solution. This required infrastructural development and, in retrospect, I felt ridiculous for not having realized sooner how far from feasible this would be. I took for granted my own grant-writing skills as a PhD-holding skilled writer with multiple successful grants to my name. I was working with a team of partners for whom English was not a native language and who had high-school educations at best.

This was revealing on many levels. My colleague on this grant and I reached out to the granting organization we had been working with and began explaining our ideas to them. This organization had lamented in the past the extremely low number of grant proposals they had received from the Pacific Islands, and we were now in a position to explain to them why this was the case. We had also come to realize that they would need to help with capacity-building in multiple areas such as grant-writing, budgeting, program management, public relations, and all-around professionalization, if they wanted their grantees to have success in submitting and competing with others globally for grants. This granting organization listened to us and began implementing such an all-around professionalization program. My colleague (on the grant and in this work) was invited to play a leadership role. For me, this has been a tangible outcome with significant positive effects on people's lives, resulting directly from my experience in the field – even if at the moment of discovery, it was a painful thing to realize because it felt like a failure. Yet we were able to turn it around and make a positive difference by not being afraid to admit to our project sponsors what we had learned and to suggest ways we could all move forward.

6.3 Developing the SolarsSPELL Digital Library

During the year that we started working with the two Peace Corps volunteers in the Chuuk Lagoon (2013), it became increasingly clear that building out the infrastructure for the multi-island connectivity project in Chuuk was not going to happen anytime soon. I decided to focus on how we could get educational content out to as many schools as inexpensively and as quickly as possible. Thus, the idea was born for a ruggedized, portable, self-powered digital library that would be as simple and straightforward as possible: the SPELL – Solar Powered Educational Learning Library.

Though I was making the transition to a new university, I was still able to offer project-focused, team-based, hands-on classes. In early 2015, I challenged my students to build something that would fit in the backpack of a Peace Corps volunteer, be waterproof, resistant to the elements, and would just plain work. We already had the website and the skeleton of an information repository that we began to consider as a library. We asked ourselves what would be useful in a library to a student or teacher in the Pacific Islands.

This time, I set up a conference call with the Peace Corps staff in the Federated States of Micronesia (FSM). They had already heard about the technology-in-the-schools work we had been doing in Chuuk over the prior 3 years and they agreed to work together with us. They invited us to come and distribute these libraries as well as conduct training on them to their incoming cohort of volunteers in August of 2015. Suddenly we had a deadline and a project to deliver. This motivates students enormously.

I pressed my luck and reached out to the Peace Corps country director in Vanuatu and found another interested party; however, he encouraged me to come to the country before their September training to meet other potential partners as well as the Peace Corps staff. Three short days before I was scheduled to leave for Vanuatu during my school's spring break, Cyclone Pam, a Category 5 storm, devastated the country. I did not make the trip. Less than a month later, a typhoon hit both Chuuk and Yap states within FSM, leaving significant destruction in its wake. It became crystal clear to my team how important it was to include content on climate change and the environment.

So, another hands-on project-based course started and I challenged this new group of students. This concept of a digital library needed to become real. Students would be designing, developing, and deploying it all within a few months' time. I always launch a course with a background overview of the project. I explain its history and its progress thus far. This is so students know that they are contributing to a real-world project that is making a difference in people's lives and that they are continuing the work of many others. However, I also launched this particular class with a design challenge. The students had a weekend of working in randomly-assigned small groups (and thus of getting to know each other) to come up with a design for this library according to the list of specifications.

The following week they presented the designs to the class. We had a winning design, but there were some ideas from another group that were also eventually incorporated. Then the students defined the smaller sub-teams that they believed would best meet the needs of the project and then individually joined these groups according to their own interests and skills areas. And once again, they delivered! Not without challenges along the way, some of which are detailed below.

We deployed our first round of 50 SolarSPELL digital libraries with Peace Corps volunteers in FSM and Vanuatu right on schedule in the summer, following the course. Figure 8.4 depicts what these libraries look like while Figure 8.5 was taken at the completion of training with Peace Corps volunteers in FSM in August 2015. The subsequent summer, in 2016, following another course and with additional content and an improved website for the library, we deployed with Peace Corps volunteers in Samoa and again in FSM.

The feedback so far has been positive. There are no hardware challenges that have been reported from the field. Mainly, we are receiving requests for additional content, which we are working to meet. Among other feedback we have received is that the volunteers are extremely grateful for the climate change content since there is no curriculum in the schools on this subject, yet, this urgent topic is extremely difficult to describe or convey without some type of teaching tools. Another heartwarming thing to hear was how the locals, whether from Vanuatu, FSM, or Samoa, love learning about their Pacific Island 'neighbors' whom they had not met or known about before these libraries arrived. We have strived to include as much local content as we can find, and it is quite gratifying to know how it is appreciated.

Fig. 8.4 SPELL: the Solar Powered Educational Learning Library

Fig. 8.5 After the first team training with the Peace Corps volunteers in the Federated States of Micronesia

As stated above, working with Peace Corps volunteers is neither a perfect nor complete solution. We will not reach all schools. There are always some volunteers who do not complete their service and a few volunteers have reported not using the library at all. While it is still too soon to report on any outcomes that are more than anecdotal, the data are being gathered. The majority of volunteers we have heard from are finding the library to be a useful tool, and most are training their teacher-counterparts, students, and others on how to use it. Some are also making the SolarSPELL available to community members to use. This is a work-in-progress, but I believe it is the project that shows the most potential to have a positive impact on schools in the Pacific Islands because it comprises (sometimes even painful) lessons learned over years of earnest attempts to understand, learn, and improve.

7 Discussion of Pedagogical Lessons Learned or Best Practices

Hands-on project-based courses are not necessarily straightforward or simple to teach or predictable in how they will turn out. It takes a great deal of bravery, adjustment, and time for a teacher to transform from 'the sage on the stage' to more of a coach or learning facilitator. It is difficult to transition from the expert with all the answers to just another member of a team in which everyone has valuable contributions to make and in which project failure is eminent without the expertise and input of others. Having led these types of classes over the course of 6 years now, I have observed a few best practices, even if they came to my realization by accident or through trial and error. What follows are some observations.

The more the students can get to know each other, the more that they trust each other. And the more students see each other (and me) working very hard on the project, the more they feel confident that this project and their work will make a difference in the real world. It helps to *ratchet up* the stakes by creating real deadlines with actual in-field partners to whom students feel beholden to deliver. Thus, the more students I can bring along with me on the deployments and into the field, the better. This also, according to students, provides a life-changing experience for them.

I would like to mention that I work with an engineer on all of these projects and the technology we are using is simple or simplified. I know that the technology works or can be made to work so I do not lose sleep over whether the students will be able to successfully deliver to a client. Students don't necessarily know this, however. Besides the technology (which anyone in this field will tell you is the easy part), everything else is an unknown and is up for grabs. For example, this means that I never know what the design will look like that the students come up with, and this applied to the large-scale direct-current solar system, the solar-computer-lab-in-a-box, as well as for the solar digital library and the websites, videos, and all accompanying materials.

I could not have predicted the difficulty in working with ministries/departments of education. I could not know in advance that Peace Corps volunteers would represent the perfect match for a solar digital library, nor the myriad of real-life challenges (including typhoons, expired passports, and tarantulas, to name but a very few) that sprang up along the way and most certainly on-site. This is where the adventure comes in, where the challenges arise, and why it helps to know that your technology should just work – so that you are prepared for the unexpected to happen throughout the rest of the endeavor.

Once the course starts, I try to organize activities early on that will allow the students to get to know each other socially. This builds friendships, interpersonal trust, team spirit, and I believe increases the level and quality of work the students put into the class.

Each class brings new challenges because each team is working on a new project or on a new facet of the project. The most memorable team experience, in my mind, was a class that was divided up into four teams. All of these teams had clear failure points during the course. The design team realized we could not afford the design we had chosen. The marketing team realized that the external website was too complicated. The content team realized that the internal website was not going to work. Not the least, the technology team realized that the server software they had chosen was too 'heavy.' In meeting with each of these groups, I listened to the challenges and made suggestions for how they could move forward. Bravely, three out of four of these teams chose to just start over, take a different direction, and thus redouble their efforts in order to reach the finish line. They had discovered what did not work and now had a clean slate for the new path they would blaze. The fourth group, unfortunately, decided to stay the course on the dead-end route they had chosen and by the end of the class had nothing successful to show for their work (in the end, we had no external website for the project at the end of the class). Never had it been so clear and pronounced to me how effective a learning tool failure can be, if we choose to learn from it and are brave enough to set out on a different path. It also became clear how brilliantly hands-on, project-based courses like these can provide such opportunities to students. I tell this story to the students in my subsequent classes, but it remains merely an illustration. They need to experience failure firsthand.

In fact, this experience was enlightening for me in many ways. I began to see my journey through these various attempts to bring this new technology to schools as a series of failures, from which I have learned, pivoted, and done better the next time. Pedagogically speaking, each course has also provided an opportunity to try new approaches and continually strive for improvement.

Failure is one of the best ways to learn, but only if we actually take the time to learn from it. Creating the time for reflection in order for that to happen is essential but it is extremely difficult to do (or to find time for). It is also important to recognize that nearly nothing happens perfectly the first time, and there's always room for improvement.

This type of teaching-and-learning approach also channels students' (ever increasing) desire to go out and make a change in the world. I sense a frustration among my students with endless book-based learning, especially when the topics are focused on things that they feel they have no control over or are happening somewhere else across the globe. What does the average student feel that they can do about wars, inequality, and poverty around the world? Not much! Therefore, activities such as project-based courses with real-world application give students the ability to contribute tangibly to making (and being a part of) positive change in the world (Benjamin-Alvarado, 2015).

Still, it can be scary. Not knowing the outcome of a class or how things will go in the field can be disconcerting for a faculty member. This type of work is also extremely time-consuming, and those who go into it rarely recognize this aspect initially. It takes far more time outside the classroom to build the relationships necessary to carry out projects like these than it does teaching traditional classes. This is exponentially more difficult when the relationships are with people on the other side of the globe. It takes proposing your idea, sending and answering countless emails, and having numerous Skype calls. It also requires being mindful of the dual focus on satisfying both home university and in-the-field partners.

This type of work also requires faculty to develop various new skills through activities such as fundraising, awareness-raising, and public relations. It involves designing a project for a semester or quarter that could feasibly be accomplished in that timeframe based upon the number of students who sign up and the skills present in that class. Plus, every time such a class is offered, it requires significant preparation time.

A person who does not do this kind of in-field work will have difficulty understanding how much outside work is involved in making it happen. Most universities do not recognize or value this type of work in the same way as traditional teaching, research, and writing. Most universities do not recognize the outsize and life-changing impact it can have on the students. Thus, at most universities, such work is not rewarded, it does not count toward tenure, and there are no accommodating reductions in class sizes. This is a shame, but the more faculty who successfully do this kind of work, and the more students whose lives are impacted by it, hopefully, the more universities will recognize its importance.

8 Conclusion

Bringing our work and ourselves out into the field carries both risk and reward. We might fail faster but we learn more intensely. We thus have the propensity to do things better and faster as well. I work on improving the technology to go out into the field because the people in these classrooms deserve the best, most appropriate technology that will simply work. But the real work is not mainly about the technology. It is about the people. Technology may draw the students to the project, but particularly once they get into the field, students realize that the true challenges have less to do with the technology than with the political and socio-cultural realities facing teachers in resource-constrained circumstances.

It takes a great deal of bravery and faith and optimism to launch into becoming a practitioner, leaving the relative comfort of one's desk chair. I was initially inspired to make the leap out of frustration from the fact that writing papers alone was not having a sufficient effect in the real world. However, it also can be daunting and a weighty feeling to know that you are directly responsible for the effect your work is having on actual people out in the world. On the other hand, we face the same anxieties when we think about the direct effect we have when we teach students.

Nothing has been more rewarding for me as a human being or as an educator than this type of work. I learn more with each trip than I ever possibly could by sitting at my desk and doing the research. What is more, the learning becomes real since I am experiencing it firsthand (and sometimes painfully). I have become both more passionate and (I believe) more effective. My work has become more rewarding, revealing, inspiring, and life-changing for the students – and is not that one of our greatest motivators as educators?

Nothing in my career as an academic has been more challenging or rewarding; nothing has been scarier or produced such a sense of pride and accomplishment than becoming a practitioner of what I preach.

References

Anderson, L. (2006). Analytic autoethnography. *Journal of Contemporary Ethnography, 35*(4), 373–395.

Bayard de Volo, L., & Schatz, E. (2004). From the inside out: Ethnographic methods in political research. *PS: Political Science and Politics, 37*(2), 267–271.

Benjamin-Alvarado, J. (2015). Internationalizing "engaged" learning: Enhancing travel study in Cuba. *Journal of Political Science Education, 11*, 483–493.

Brewer, J. D. (2000). *Ethnography*. Philadelphia, PA: Open University Press.

Central Intelligence Agency (CIA). (1999). *Map of the Federated States of Micronesia* [Online image]. Retrieved from http://www.lib.utexas.edu/maps/micronesia.html

Central Intelligence Agency (CIA). (2017). *The world factbook. Retrieved from https://www.cia.gov/library/publications/the-world-factbook/geos/fm.html*

Gerring, J. (2004). What is a case study and what is it good for? *American Political Science Review, 98*(2), 341–354.

Hosman, L. (2014). International service-learning partnerships: Lessons of successes and challenges that defy simple categorization. *Partnerships: A Journal of Service Learning and Civic Engagement, 5*(1), 24–50.

Hosman, L., & Cvetanoska, M. (2013). Teachers, technology, & training: Combining theory with Macedonia's experience. *International Journal of Education and Development using ICT, 9*(3), 28–49.

Houghton, C., Casey, D., Shaw, D., & Murphy, K. (2013). Rigour in qualitative case-study research. *Nurse Researcher, 20*(4), 12–17.

Interamerican Development Bank (IDB). (2007). *Technical cooperation profile*. Retrieved from http://idbdocs.iadb.org/wsdocs/getdocument.aspx?docnum=1223629

Kapp, E. (2009). Improving student teamwork in a collaborative project-based course. *College Teaching, 57*, 139–143.

Kozma, R. B., & Voogt, J. (2003). *Technology, innovation, and educational change: A global perspective* [A Report of the Second Information Technology in Education Study, Module 2]. Eugene, OR: International Society for Technology in Education.

Maher, M. M. (2003). Individual beliefs and cultural immersion in service learning: Examination of a reflection process. *The Journal of Experiential Education, 26*, 88–96.

Marcus, G. E. (1995). Ethnography in/of the world system: The emergence of multi-sited ethnography. *Annual Review of Anthropology, 24*, 95–117.

Marlin-Bennett, R. (2002). Linking experiential and classroom education: Lessons learned from the American University-Amnesty International USA Summer Institute on Human Rights. *International Studies Perspectives, 3*, 384–395.

Mumtaz, S. (2000). Factors affecting teachers' use of information and communications technology: A review of the literature. *Journal of Information Technology for Teacher Education, 9*(3), 319–342.

Patterson, M., & Renwick Monroe, K. (1998). Narrative in political science. *American Review of Political Science, 1*, 315–331.

Peterson, T. H. (2009). Engaged scholarship: Reflections and research on the pedagogy of social change. *Teaching in Higher Education, 14*, 541–552.

Pierides, D. (2010). Multi-sited ethnography and the field of educational research. *Critical Studies in Education, 51*(2), 179–195.

Thomson, I. (2015). *Overview of ICTs in Pacific Education* [Presentation to ICT Officials at the International Telecommunications Union]. Retrieved from https://www.itu.int/en/ITU-D/Regional-Presence/AsiaPacific/Documents/Events/2015/June-Pacific-Ministerial-Meeting/USP.pdf

Tomlinson-Clarke, S. M., & Clarke, D. (2010). Culturally focused community-centered service-learning: An international cultural immersion experience. *Journal of Multicultural Counseling and Development, 38*, 166–175.

Wall, S. (2006). An autoethnography on learning about autoethnography. *International Journal of Qualitative Methods, 5*(2), 146–160.

Weeden, L. (2010). Reflections on ethnographic work in political science. *Annual Review of Political Science, 13*, 255–272.

Williams, C. J. (2009). *E-Granary, digital literacy, and the identities of Ugandan students* (Unpublished doctoral dissertation). University of British Columbia Library, Vancouver, BC. doi: 10.14288/1.0067483

Wozney, L., Venkatesh, V., & Abrami, P. C. (2006). Implementing computer technologies: Teachers' perceptions and practices. *Journal of Technology and Teacher Education, 14*(1), 173–207.

Yin, R. K. (2003). *Case study research: Design and methods* (3rd ed.). London, England: Sage Publications.

Author Information

Laura Hosman
Arizona State University
Interdisciplinary B256K, 1120 S. Cady Mall, PO Box 875603, Tempe, AZ 85287-5603

Complete mailing address: Arizona State University, Interdisciplinary B256K, 1120 S. Cady Mall, PO Box 875603, Tempe, AZ 85287-5603
Telephone number: 480-727-6616
Email address: laurahosman@gmail.com
Website: https://ict4dviewsfromthefield.wordpress.com

Laura Hosman is Assistant Professor at Arizona State University, holding a joint appointment in the School for the Future of Innovation in Society and in The Polytechnic School. With an emphasis on action-oriented, in-the-field work, her research focuses on the role of information and communications technology (ICT) in developing countries, particularly in terms of its potential effects on socio-cultural factors, human development, and economic growth. Presently, she focuses on ICT-in-education projects and brings her passion for experiential learning to the classroom by leading real-world-focused, project-based courses that have seen student-built technology deployed in schools in Haiti, Vanuatu, Micronesia, Samoa, and more. She has held prior academic positions at California Polytechnic State University, Illinois Institute of Technology, the University of California, Berkeley, and the University of Southern California (USC). She earned an MA in International Relations as a Fulbright Scholar at the University of Amsterdam. She also holds an MA in Economics and a Ph.D. in Political Economy and Public Policy, both from the University of Southern California (USC).

Errata to: ICT-Supported Innovations in Small Countries and Developing Regions: Perspectives and Recommendations for International Education

Ian A. Lubin

Errata to:
I.A. Lubin (ed.), *ICT-Supported Innovations in Small Countries and Developing Regions*, Educational Communications and Technology: Issues and Innovations, https://doi.org/10.1007/978-3-319-67657-9

The author information available at the end of the chapter has been updated for the below chapters.

In Chapter 1, the mailing address and the author biography has been updated in page 26.

In Chapter 6, the mailing address of the author Tian Belawati has been updated in page 137.

In Chapter 7, the mailing address of the author Jeremiah H. Kalir has been updated in page 167.

The updated online versions of these chapters can be found at
https://doi.org/10.1007/978-3-319-67657-9_1
https://doi.org/10.1007/978-3-319-67657-9_6
https://doi.org/10.1007/978-3-319-67657-9_7
https://doi.org/10.1007/978-3-319-67657-9

I.A. Lubin (ed.), *ICT-Supported Innovations in Small Countries and Developing Regions*, Educational Communications and Technology: Issues and Innovations,
https://doi.org/10.1007/978-3-319-67657-9_9

Index

I.A. Lubin (ed.), *ICT-Supported Innovations in Small Countries and Developing Regions*, Educational Communications and Technology: Issues and Innovations,
DOI 10.1007/978-3-319-67657-9

MIX
Papier aus verantwortungsvollen Quellen
Paper from responsible sources
FSC® C105338

If you have any concerns about our products,
you can contact us on
ProductSafety@springernature.com

In case Publisher is established outside the EU,
the EU authorized representative is:
Springer Nature Customer Service Center GmbH
Europaplatz 3, 69115 Heidelberg, Germany

Printed by Libri Plureos GmbH
in Hamburg, Germany